Lecture Notes in Physics

For information about Vols. 1–151, please contact your bookseller or Springer-Verlag.

Lecture Notes in Physics

Edited by H. Araki, Kyoto, J. Ehlers, München, K. Hepp, Zürich
R. Kippenhahn, München, H. A. Weidenmüller, Heidelberg
and J. Zittartz, Köln
Managing Editor: W. Beiglböck, Heidelberg

227

Jean-Pierre Eckmann
Peter Wittwer

Computer Methods and Borel Summability Applied to Feigenbaum's Equation

Springer-Verlag
Berlin Heidelberg GmbH

Authors

Jean-Pierre Eckmann
Département de Physique Théorique, Université de Genève
32, boulevard d'Yvoy, CH-1211 Genève

Peter Wittwer
Courant Institute, New York University
251 Mercer Street, New York, NY 10012, USA

ISBN 978-3-540-15215-6   ISBN 978-3-540-39359-7 (eBook)
DOI 10.1007/978-3-540-39359-7

Originally published by Springer-Verlag Berlin Heidelberg New York Tokyo in 1985

2153/3140-543210

Table of Contents

Preface

This book is an outgrowth of a seemingly trivial question, namely : what happens with the Feigenbaum universality when one restricts attention to functions of $|x|^{2N}$ (instead of x^2), with N very large ? Closer numerical investigation [26,29,35,40] reveals that this question is hard, and that some aspects of the problem must become singular in the limit as $N \to \infty$. In this work, we study this singular limit, and sketch its relation to the case of finite, but large N.

The proof of the existence of this singular limit is a difficult problem in which several new ideas are needed. We believe that some of these ideas could serve as a guideline for other problems, as we shall indicate below. The main themes are :

1) Feigenbaum universality [19,20]

2) Ecalle's theory of resurgent functions [12,13]
   (Alternatively, this can be viewed as a constructive version of renormalization group problems in marginal situations.)

3) Constructive aspects of Borel summation [33]

4) Techniques for computer assisted proofs [15,16,32]

Each of these problems is addressed in a largely self-contained section, so that a reader can learn about one of the above subjects without reading the whole book.

We shall next outline the problem and describe the four subjects above.

1)   Feigenbaum universality

Denote by I the interval [-1,1] and consider, for fixed N > 1/2, the one parameter family of maps from I to I given by

$$\mu \to \psi_\mu(x) \equiv 1 - \mu|x|^{2N} ,$$

with $\mu \in [0,2]$. Clearly ψ_μ maps I into itself and ψ_μ has a critical point at x = 0. For each μ, one views ψ_μ as a dynamical system, i.e. one studies the sequences of iterates $\{x_n\}$, $x_n = \psi_\mu(x_{n-1})$, with $x_o \in I$ given. It can be shown that for every N > 1/2 there is a smallest $\mu_m^{(N)}$ for which the sequence $\{x_n\}$ is periodic with period 2^m, when $x_o = 0$. One says that $\psi_{\mu_m^{(N)}}$ has a superstable period of length 2^m.

In studying the case of N = 1 numerically, Feigenbaum [19] found the following striking universality : the sequence $\mu_m = \mu_m^{(1)}$ has a limit μ_∞ ($\mu_\infty \neq 0,2$) and for large m, $\mu_\infty - \mu_m$ is asymptotically proportional to

$$\text{const.} \cdot \delta^{-m}$$

where $\delta = \delta_F = 4.669...$ is apparently the same for several different one-parameter families, such as the one above, or the family $\cos \mu x$. (Note that the above property does not depend on a differentiable change of coordinates in μ, provided the derivative at μ_∞ is not zero.)

In marked contrast, numerical experiments [19] and later analytic results [9] show that δ does depend on N, i.e. on the nature of the critical point of the ψ_μ, $\delta \equiv \delta_N$.

The number $\delta = \delta_N$ has a functional analytic origin, as follows. One considers, in the space of analytic functions of $|x|^{2N}$ the non-linear operator

$$N : \psi(x) \rightarrow \frac{1}{\lambda_\psi} \psi(\psi(\lambda_\psi x))$$

($\lambda_\psi = \psi(1)$). One looks for a fixed point φ_N of N, i.e. a function satisfying

$$\varphi_N(x) = \frac{1}{\lambda_N} \varphi_N(\varphi_N(\lambda_N x)) \quad (\equiv N \varphi_N(x)), \quad \varphi_N(0) = 1, \tag{1}$$

where φ_N is of the form

$$\varphi_N(x) = f_N(|x|^{2N}),$$

with f_N analytic. The number δ_N is then identified with the (only) eigenvalue of the tangent map DN (at φ_N and acting on a suitable Banach space) which lies outside the unit disk.

The case $N = 1$, i.e. the situation where $\psi_\mu(x)$ is of the form $\psi_\mu(x) = 1 - \mu x^2$, turned out to have ramifications way beyond the simple theory of maps of the interval. It was in fact shown by Collet, Eckmann and Koch [7] that also for families of maps on $\mathbb{R}^m$ the constant $\delta_F = 4.669...$ will appear naturally. This clearly opened the field to dynamical systems describing physical situations. In fact, many experiments have been performed which show the characteristic accumulation of period doublings, and in some cases the accumulation rate has been reproduced to within experimental limits.

It should also be mentioned that the Feigenbaum universality is one of the most clear-cut examples of the "renormalization group method". In contrast to statistical mechanics there is no approximation necessary to find the renormalization group action (it is the operator N) and to show the existence of a non-trivial fixed point. (Even in $\mathbb{R}^m$, the action of N is a sort of "decimation" only in the time direction, and things are exact because time is one-dimensional.)

The reader might wonder why there is still some interest in this problem, since everything falls so beautifully into place. From our point of view there are two reasons. First, since the renormalization group is relatively simple, without being trivial, it is a good testing ground for general ideas about renormalization groups. The limit we study here is the one in which the fixed point has a scaling going to 1 (in fact, -1), and such situations are of interest e.g. in the XY-model. Second, the actual proofs of the existence of a fixed-point Collet, Eckmann, Lanford [9], Campanino, Epstein, Ruelle [3], Lanford [32] all suffer from a lack of final insight; see e.g. Lanford [31] for a review of this problem. By looking at the problem from a different perspective, one can hope to understand it better. Here we consider the limit $N \to \infty$; another limit is increasing the number of iterations,

$N_p \varphi(x) = \frac{1}{\lambda} \varphi^p(\lambda x)$, where φ^p is the p-fold iteration of φ; see Eckmann, Epstein, Wittwer [14].

We study here in particular the limit $N = \infty$ (Section 5) and sketch the proof for the case $N < \infty$, N large (Section 6). It should be pointed out that several authors have done numerical experiments on this problem [26,29]. But it is very hard to do reliable computer experiments because the function $1 - \mu|x|^{2N}$ is difficult to evaluate for large N. Also, the convergence of the bifurcation points starts out to look first like corresponding to a large value of $\delta = \delta_N$ and only very high bifurcations which are numerically inaccessible are related to the true δ_N. (See Fig. 1 below.) Therefore computer experiments and perturbative renormalization group calculations all seem to indicate that $\delta_N \to \infty$ as $N \to \infty$ (see however [40]), but we conjecture that in fact

$$\lim_{N\to\infty} \delta_N = \delta_\infty \ , \quad \text{with} \quad \delta_\infty \ \underline{\text{finite}} \ . (\delta_\infty \gtrsim 30).$$

Thus, we are looking for solutions φ_N of the form $\varphi_N(x) = f_N(|x|^{2N})$, and we shall require $f_N(z)$ to be analytic in a complex neighborhood of the real segment $[0,1]$. Then f_N must satisfy

$$f_N(z) = \frac{1}{\lambda_N} f_N([f_N(|\lambda_N|^{2N}z)]^{2N}) , \tag{2}$$

$$f_N(0) = 1 \ , \quad f_N(|\lambda_N|^{2N}x) > 0 \quad \text{for} \quad x \in [0,1]$$

(necessarily, $\lambda_N = f_N(1)$).

We treat this equation as a perturbation problem around the limit $N = \infty$, which itself again is a non-trivial equation. The limit of the f_N is singular as $N \to \infty$ (f_∞ is a step function) but the functions $g_N(z) = [f_N(|\lambda_N|^{2N}z)]^{2N}$ and the number

X

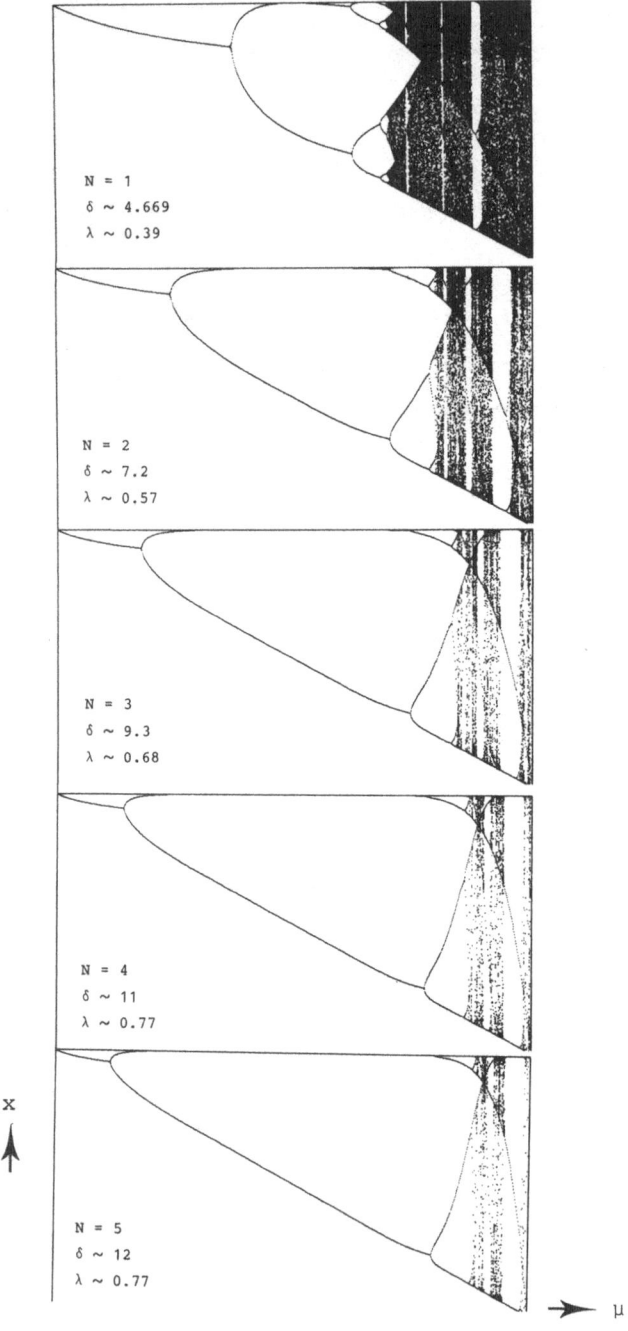

N = 1
$\delta \sim 4.669$
$\lambda \sim 0.39$

N = 2
$\delta \sim 7.2$
$\lambda \sim 0.57$

N = 3
$\delta \sim 9.3$
$\lambda \sim 0.68$

N = 4
$\delta \sim 11$
$\lambda \sim 0.77$

x

N = 5
$\delta \sim 12$
$\lambda \sim 0.77$

μ

Fig. 1. Bifurcation diagrams for the families $1-\mu x^{2N}$ for several values of N.

$\tau_N = |\lambda_N|^{2N}$ will be seen to have nice and non-trivial limits, g_∞ and τ_∞. We shall not formulate this result now, but refer to the introduction.

2) Ecalle's theory of resurgent functions

In the course of the construction of g_∞ and τ_∞ we shall be naturally led to consider the following system of equations :

$$h_\infty(g_\infty(g_\infty(z))) - h_\infty(z) = -2 , \tag{3}$$

$$g_\infty(z) = \exp(-\log\tau_\infty \cdot h_\infty(\tau_\infty z)) , \tag{4}$$

where h_∞, x_∞ and τ_∞ are chosen so that

$$h_\infty(0) = 0, \quad g_\infty(x_\infty) = x_\infty , \quad g_\infty'(x_\infty) = -1 .$$

Of special interest for this preface is the equation (3). This equation (we view g_∞ as given and h_∞ as unknown) is of Abel's type. The difficulty with this equation comes from its "marginality" : The derivative of g_∞ at its fixed point x_∞ is -1. Equations of this type have been extensively studied in Ecalle's work[12,13]. The main conclusion for our purpose is the following : because $g_\infty'(x_\infty) = -1$, h_∞ cannot be analytic at x_∞. Nevertheless, h_∞ is totally determined by its asymptotic expansion at x_∞ and in addition it is a Borel summable function. The asymptotic expansion of h_∞ can be calculated from the Taylor series expansion of g_∞ at x_∞. We believe that the general conclusion : "$g_\infty'(x_\infty) = e^{i\pi(p/q)}$ implies Borel summability" should play an important role in many physics problems with marginal eigenvalues. Examples could be suggested by the ϕ^4 field theory in 4 dimen-

sions [37,38],by the WKB method [41], or by special models of statistical mechanics [11]. We hope that our work will help to stim-
ulate a deeper understanding of the relation between marginality
and Borel summability.

3)   Constructive aspects of Borel summability

    A function f, which is analytic in a sector
$\{z \mid |z| < R, \ |\arg z| < \frac{\pi}{2} + \sigma, \sigma > 0\}$ is called Borel summable if
its Taylor coefficients satisfy

$$|f_n| \leq n! \ c^n A$$

and

$$\left| f(z) - \sum_{k=0}^{n-1} f_k z^k \right| \leq n! \ c^n |z^n| A$$

on the sector. In general, the Taylor expansion of Borel summable
functions diverges, but by a well-known theorem of Borel (see e.g.
[42]) the function f is uniquely determined by its coefficients
f_n.   It is also known (see e.g. [43]) that there are effective
procedures to actually compute f from f_n. We shall use here
the method described by Loeffel [33], and shall give the necessary
estimates which occur in the case at hand. As far as we know, this
is the first example of a constructive evaluation of a Borel sum
with rigorous error bounds.

    Borel summable functions occur in many physical problems,
not all related to equations of Ecalle's type. The following list
is certainly incomplete, but illustrates how widespread applications are : the quartic oscillator [41] and anharmonic oscillators

[44], $\varphi_{2,3}^4$ quantum field theories [45,46], the Stark effect [47].

There are also indications of Borel summability for other problems :

the ε-expansion in the renormalization group [39] and in the

hierarchical model of Dyson [48].

4) Techniques for computer-assisted proofs

The proofs of most of our theorems are based on estimates

done with the help of a computer. Although this method, which was

pioneered by Lanford [32] has already been documented in some de-

tail [16], we believe that a broader knowledge of this method is

to be encouraged and that in the long run parts of mathematical

physics should profit from the availability of a powerful auto-

matic analytic tool.

The subroutine package described here handless functions in

one variable, whereas in Eckmann, Koch, Wittwer [16], functions in

two variables where considered. The subroutines have been improved,

some estimates are more general (e.g. composition of functions) and

the notation has been streamlined. A Pascal version of the program

is also available. With such packages, it should be relatively

straightforward to implement again Lanford's proof of the existence

of Feigenbaum's fixed point for the case $N = 1$.

It is typical for problems in mathematical physics that they
are somehow given from the outside. As an example, Coulomb scattering
plays a more fundamental role than the scattering theory of short-
range potentials, although the latter are much easier to handle.
It is also evident that the degree of difficulty of a problem
which is given from the outside cannot really be lowered. This
is the curse and the challenge of mathematical physics. The tool
of mathematically rigorous computer-assisted bounds seems therefore very adequate and it should be helpful in problems with a
naturally high degree of computational complexity and arithmetic
work. We shall explain the method again in detail and document the
proof by furnishing the listing of the program. The general routines of the program should in the future become somewhat of a
generally accepted tool, which can be used on every desk-calculator
or computer. For the moment it seems that the tool is mostly useful
for problems involving the composition of functions, but one can
hope that other domains of analysis could be implemented in a similar spirit.

1. Introduction

The main purpose of this work is to prove the existence of
a class of solutions of Feigenbaum's functional equation

$$\varphi(x) = \frac{1}{\lambda}\varphi(\varphi(\lambda x)) ,$$

$$(1.1)$$

$$\varphi(0) = 1 ,$$

where φ is a map of the interval $I = [-1,1]$ into itself. The
functions we want to consider are to be increasing on $[-1,0]$, and
decreasing on $[0,1]$, so that the corresponding dynamical system
is a two to one map. Solutions of this form do not seem to fall
into the class of tabulated functions and must therefore be found
by analytic methods. The second equation of (1.1) implies $\lambda = \varphi(1)$,
so that the problem is well-posed. However we still have the freedom of choosing the function space in which we want to solve the
equation. For example, to each $N > 1/2$ there corresponds
the problem of finding a solution of (1.1) behaving, for small $|x|$
like $1 - \text{const.}|x|^{2N}$. For $2N = 1 + \varepsilon$ and ε sufficiently
small, the problem has been fully solved by Collet, Eckmann and
Lanford [9] , who show that there is a ε-dependent solution
$\varphi_\varepsilon(x) = f_\varepsilon(|x|^{1+\varepsilon})$, f_ε analytic. The problem $N = 1$ has been
solved by Campanino, Epstein and Ruelle [3,4] and by Lanford
[32], who makes essential use of mathematically rigorous computer
methods. In our work we show - using again such methods - that
for N sufficiently large the problem has a N-dependent solution
$\varphi_N(x) = f_N(|x|^{2N})$, where f_N is analytic on a domain D_N as
indicated in Fig. 1.1 below and solves there the equation

$$f_N(z) = \frac{1}{\lambda_N} f_N([f_N(|\lambda_N|^{2N} z)]^{2N}) ,$$

$$(1.2)$$

$$f_N(0) = 1 .$$

Our interest in the large N limit is partially motivated by the fact that the difficulties in solving the problem are "symptomatic" for a class of conjugation problems in mathematical physics. See e.g. the "asymptotic freedom" in field theories, Francoise's [49] analysis of scattering theory or Voros' [41] treatment of the quartic oscillator. These problems have in common the presence of a non-hyperbolic structure, and an associated decomposition of analyticity domains. The saving grace in those problems seems to be the theory of "resurgent functions" (Ecalle [13]), in which one realizes a striking recovery of analyticity through the use of Laplace transforms and Borel summability. This paper can be viewed as a detailed analysis of such a problem, combined with rigorous computer estimates.

We shall view the problem (1.2) for large N as a "perturbation" of the case N = ∞. This limiting case will turn out to be singular, because D_N splits in the limit into two pieces with only a point $x_\infty \in [0,1]$ in common, and f_N has a limit as N goes to ∞, which is singular at x_∞. In this work, we discuss first the limiting case N = ∞ (Section 5), and consider then perturbation theory, starting from N = ∞ (Section 6).

We are now confronted with the question of which equation is to replace (1.2) in the limit where N goes to ∞. Once we have decided which should be the correct equation, we shall show that it has a solution (Theorem 5.4.2). In Section 6, we shall then derive solutions of (1.2) for N near ∞ by perturbation theory, and this will, a posteriori, justify the correctness of our choice of the limit equation.

Our basic intuitive ideas are as follows : we study, instead of (1.2), the system (see Fig. 1.1)

$$f_N(z) = \frac{1}{\lambda_N} f_N(g_N(z)) , \tag{1.3}$$

$$g_N(z) = [f_N(\tau_N z)]^{2N} , \tag{1.4}$$

$$\tau_N = |\lambda_N|^{2N} , \tag{1.5}$$

$$f_N(0) = g_N(0) = 1 . \tag{1.6}$$

Past experience [4] seems to indicate that the functions g_N have chances of being more regular then f_N , in the limit $N \to \infty$. We are thus motivated in devising a scheme which focuses on these functions in the limit of large N. We now propose this scheme in detail, based on a few working assumptions. It will lead to a limiting equation ((1.7)-(1.10) below). Once this equation is solved, it will be possible to solve (1.3)-(1.6), which proves the correctness of our working assumptions, so that the seeming circularity of the argument disappears.

The assumptions about the limiting behaviour of the solution f_N and g_N of the system (1.3)-(1.6) are :

1) $\lim_{N \to \infty} \lambda_N = -1 ,$

2)   <u>for</u> $\tau_N = |\lambda_N|^{2N}$ <u>we have</u> $\lim_{N\to\infty} \tau_N = \tau_\infty$, $\tau_\infty \sim 0.033381$,

3)   <u>the functions</u>. g_N <u>are analytic in</u>

$D(0.78,1) = \{z\in\mathbb{C}\,|\,|z-0.78| < 1\}$ <u>and</u> $\lim_{N\to\infty} g_N = g_\infty$, <u>uniformly</u>

<u>on this domain</u> ,

4)   <u>every</u> g_N <u>has a fixed point</u> x_N <u>(cf. Fig. (1.1) and</u>

$\lim_{N\to\infty} x_N = x_\infty$, $x_\infty \sim 0.391$,

5)   <u>for every</u> $x\in[0,1]$, $\lim_{N\to\infty} f_N(x) = \text{sign}(x_\infty-x)$. <u>Furthermore the</u>

<u>functions</u> $h_N(z) = 2N/\log\tau_\infty \cdot (f_N(z)-1)$ <u>are analytic on</u>

$\tau_\infty \cdot D(0.78,1) = \{\tau_\infty z\,|\,z\in D(0.78,1)\}$ <u>and</u> $\lim_{N\to\infty} h_N = h_\infty$, <u>uniform-</u>

<u>ly on this domain.</u>

We shall see that these solutions are locally unique in an
adequate function space.

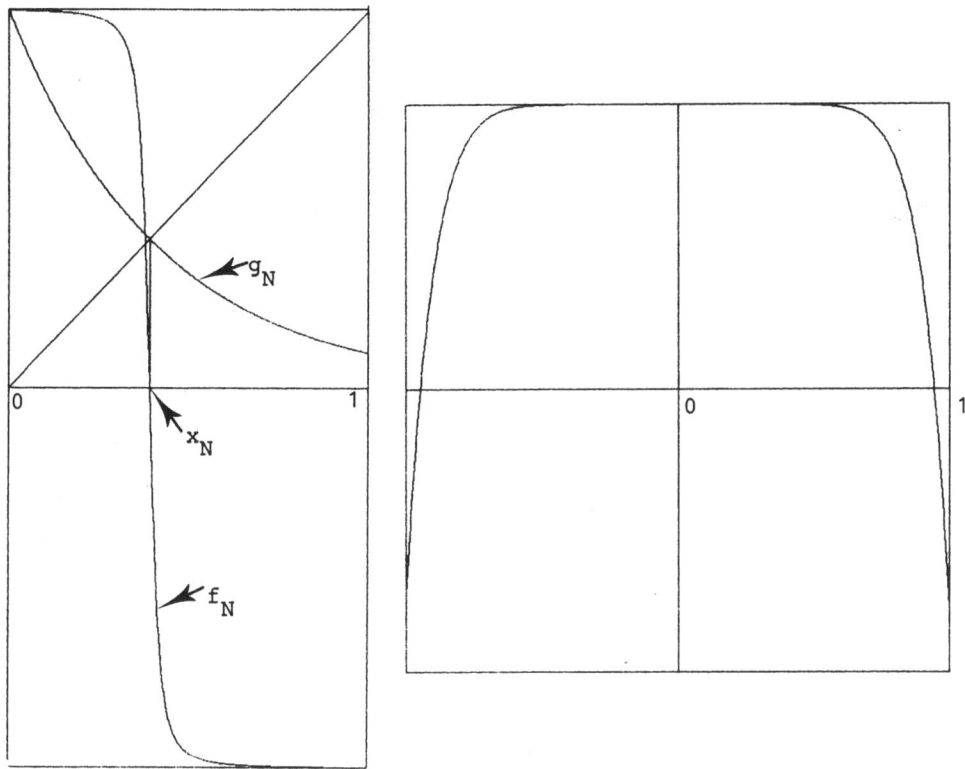

Fig. 1.1. Graphs of the functions f_N, g_N and φ_N , $N \sim 1000$.

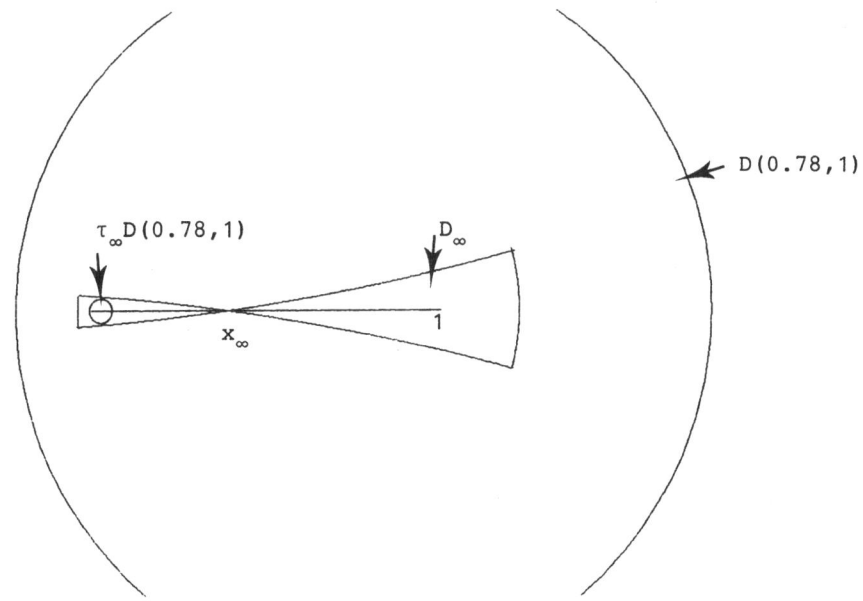

Fig. 1.2. The analyticity domains D_∞, $D(0.78,1)$, and $\tau_\infty \cdot D(0.78,1)$.

From our assumptions it follows immediately that

$$\lambda_N = -1 - \frac{\log \tau_\infty}{2N} + o(\tfrac{1}{N}) , \quad N \to \infty .$$

Approximating furthermore τ_N in (1.4) by τ_∞ one easily checks that in order to have a family of functions g_N which remains analytic on the disk $D(0.78,1)$ in the limit where N goes to infinity one has to have that the function f_N is of the form $f_N(z) = 1 + O(\tfrac{1}{N})$ for z in a compactum to the left of the singularity near x_∞. This is exactly what 5) says, and if we put in this sense

$$f_N(z) = 1 - \frac{\log \tau_\infty}{2N} h_\infty(z) + o(\tfrac{1}{N}) ,$$

we get for g_N

$$g_N(z) = [1 - \frac{\log \tau_\infty}{2N} h_\infty(\tau_\infty z) + o(\tfrac{1}{N})]^{2N}$$

$$\xrightarrow[N\to\infty]{} \exp(-\log \tau_\infty \cdot h_\infty(\tau_\infty z)) .$$

We discuss now (1.3). Let $x_N = g_N(x_N)$ be the (unique) fixed point of the function g_N in $[0,1]$ as asserted by 4). Taking the first derivative of (1.3) at x_N leads to

$$f_N'(x_N) = \frac{1}{\lambda_N} f_N'(g_N(x_N)) \cdot g_N'(x_N) = \frac{1}{\lambda_N} f_N'(x_N) \cdot g_N'(x_N)$$

and it follows that $g_N'(x_N) = \lambda_N$. By assumption we have that $\lim_{N\to\infty} \lambda_N = -1$ and $\lim_{N\to\infty} x_N = x_\infty$. Being given this local structure of g_N at its fixed point we find that it maps a compactum to the left of the singularity of f_N near x_∞ on a compactum to the right of this singularity and vice versa. Therefore, and because the analyticity domain D_N decomposes in the limit where N goes to infinity, we can not hope to find for (1.3) a reasonable limit at N = ∞. But iterating equation (1.3) once we get

$$(\frac{1}{\lambda_N})^2 f_N(g_N(g_N(z))) - f_N(z) = 0$$

and assuming for f_N the behaviour 5) we get for N → ∞ the equation

$$h_\infty(g_\infty(g_\infty(z))) - h_\infty(z) = -2$$

at least for z in a compactum to the left of x_∞, and a similar argument would lead to exactly the same equation for z in a compactum to the right of x_∞. What the domains of analyticity for h_∞ exactly look like will be seen later on.

This formal discussion motivates us to replace (1.3)-(1.6) at N = ∞ by the set of equations

$$h_\infty(g_\infty(g_\infty(z))) - h_\infty(z) = -2 , \tag{1.7}$$

$$h_\infty(0) = \beta , \tag{1.8}$$

$$g_\infty(z) = \alpha \cdot \exp(-\log\tau_\infty \cdot h_\infty(\tau_\infty z)) , \tag{1.9}$$

$$g'_\infty(x_{OO}) = -1 \quad , \quad g_\infty(x_{OO}) = x_{OO} \quad . \tag{1.10}$$

Note that τ_∞ is determined by the equation $g'_\infty(x_{OO}) = -1$, with $x_{OO} = g_\infty(x_{OO})$ the unique fixed point of g_∞ in $[0,1]$. We want to solve these equations for $\beta = 0$, $\alpha = 1$ and $x_{OO} = x_\infty$. However different choices of α and β amount only to a different normalization of the solution and hence we feel free to omit the conditions $\alpha = 1$ and $\beta = 0$ in the sequel, replacing them by $\alpha \neq 0$, $\beta \in \mathbb{R}$. In fact we shall always first fix β and afterwards choose α such that on the new scale x_{OO} turns out to be equal to a fixed number, $x_{OO} \approx 0.391$.

The main technical result of this work is the following theorem.

Theorem 1.1. Fix $x_{OO} = 0.391132999351022542^*$ and choose $\beta = 0$. There exists a real number $A \in [-.556999086, -.556999080]$, a function h_∞ analytic on $S_A^- = \{\frac{z}{1-Az} \mid |z| < .53, \ |argz - \pi| < .135\}$, a function g_∞, analytic on $D(0.78,1) = \{z \in \mathbb{C} \mid |z-0.78| < 1\}$ and real numbers α and τ_∞, $x_\infty = \alpha x_{OO} \in [0.391122, 0.391144]$, $\tau_\infty \in [0.0333808, 0.0333813]$ such that (1.7) is satisfied on S_A^-, (1.9) on $D(0.78,1)$ and such that the conditions (1.8), and (1.10) are fulfilled.

We shall see that the difficulties encountered with the study of (1.7) are similar to those found when N is finite. However, although $|\lambda_N| \neq 1$ for all finite N, we have $\lim\limits_{N \to \infty} \lambda_N = -1$ and hence a perturbation theory in N must take into account the singularities generated by the non-hyperbolicity of the limit. Thus

---

* for a precise definition of the $=$ sign, see Section 5.5.

we are faced with a perturbation theory around a limiting case which is itself singular, and this presents one of the interesting features of this work, and could serve as a guide for other situations with singular limits.

We now outline our strategy for solving (1.7)-(1.10). We reformulate this system as a fixed point problem for an operator K_{oo} which we now define. Given g, the operator K_{oo} constructs a new g as follows

i) given g solve (1.7), i.e. find h such that

$$h(g(g(z))) - h(z) = -2 ,$$

$$h(0) = \beta (= 0) ,$$

ii) find τ such that

$$x_{oo}(-\log\tau) \cdot \tau \cdot h'(\tau x_{oo}) = -1 ,$$

iii) fix $\alpha = x_{oo} \cdot \exp(\log\tau \cdot h(\tau x_{oo}))$,
iv) define $\tilde{g}(z) = \alpha \cdot \exp(-\log\tau \cdot h(\tau z))$.

The operator K_{oo} is given by $K_{oo}(g) = \tilde{g}$. It will be seen that we have reformulated the original problem in such a way that K_{oo} will turn out to be a contraction on a suitable Banach space. We shall show that

1) K_{oo} is defined on a certain class of functions,

2) there is a g_o for which, in a suitable norm,

$$\|K_{oo}(g_o) - g_o\| < \varepsilon$$

3) $\|DK_{oo,g}\| < \rho < 1$,   for $\|g - g_o\| < \beta$.

If $\varepsilon < (1-\rho)\beta$, then, by the contraction mapping principle there will be an unique fixed point in the ball in question, and hence we will have solved (1.7)-(1.10).

Lemma 1.2. If K_{oo} is defined and has a fixed point then (1.7)- (1.10) has a solution.

This is obvious from the definition of K_{oo}.

A remark concerning our formulation of the problem seems in place. Of course, there are a priori many ways of attacking the problem and our choice is only one possibility. First of all we have, guided by previous experience [16] decided to use a computer for our proof. It seems that the non linear and non local aspects of (1.9) make other - and maybe more synthetic-methods not very useful. Once we have taken the option of a computer assisted proof, we are forced to stay totally within the scope of constructive analysis. This means e.g. that existence theorems have to be fur- nished with bounds. This problem is particularly painfully felt in connection with (1.7). We are in the fortunate situation that the theory by Ecalle [13] furnishes the desired bounds, and we shall devote the next section to a detailed study of this question.

The book is organized as follows. In Section 2 we discuss in detail Ecalle's work, when applied to equations of Abel's type.

Section 3 is devoted to the method of constructive Borel summation as described by Loeffel [33]. In Section 4 we discuss the
method of computer assisted proves and present the basic set of
subroutines which are used in this method. Section 5 contains a detailed discussion of the singular limit $N = \infty$ together with a
thorough discussion of the proof, i.e. the computer program.
Finally Section 6 contains a sketch of the proof of the existence
of a solution of (1.3)-(1.6) for finite but large N.

Sections 2, 3 and 4 are each completely selfcontained and
can serve as an introduction to Ecalle's work, constructive Borel
summation and computer assisted proofs respectively.

2. Ecalle's theory of resurgent functions

2.1. Introduction

In this section we present in detail those aspects of Ecalle's theory which will be needed in later sections. We believe that this theory is of general interest, independently of the problem at hand, and it should play an increasing role in other problems with marginal eigenvalues. The present section can be read independently of the remainder of the book.

Consider as a first example the equation

$$F_o(\Gamma(z)) - F_o(z) = -2 , \qquad\qquad (2.1.1)$$

where Γ is given, and is say, of the form

$$\Gamma(z) = z - \frac{2}{\kappa} z^3 + O(z^5) ,$$

$\kappa > 0$. Suppose furthermore that Γ is analytic near $z = 0$. The task is to find F_o. For the reader who is not acquainted with this problem, let us restate some basic facts.

As a first step in solving (2.1.1) one might try to consider a solution in terms of a formal power series P_o. This does not a priori work, since for $P_o(x) = \sum_{j=0}^{\infty} P_{oj}x^j$ one would have in lowest order $P_{oo} - P_{oo} = -2$, a contradiction. Thus if (2.1.1) has a solution at all, it has to be singular at $z = 0$. In fact we shall see that F_o is of the form $F_o = F_{sing} + F_{reg}$, with

$$F_{sing}(z) = \frac{a_o}{z^2} + c_o \log z^2 + d_o z .$$

With such an ansatz for F_o one can in fact start to determine recursively the coefficients of a formal power series solution $P_{reg}(x) = \sum_{j=2}^{\infty} P_{reg,j} \cdot x^j$. Quite generally, the cancellation which forced us to introduce F_{sing} persists at higher order; we have

$$(\Gamma(z))^k - z^k = 0(z^{k+2}) , \quad k > 0$$

and as a result the series $P_{reg}(x)$ <u>diverges</u> in general and does not a priori define a function F_{reg}. The saving grace comes from two sources, both analyzed in Ecalle's work [13] :

i) The orbits of Γ are weakly attracted in sectorial domains $\mathfrak{s}^{\pm}$ to the right and left of $z = 0$. If $z \in \mathfrak{s}^{\pm} = \{z = \pm\rho e^{i\phi} \big| |\phi| < \phi_o , 0 < \rho < \rho_o\}$ for some suitable ϕ_o, ρ_o, then $\Gamma^n(z) \sim 0(n^{-1/2})$, $n \to \infty$. (This follows from $\mathrm{Re}((\frac{1}{\Gamma(z)})^2 - \frac{1}{z^2}) \geq L > 0$, see below.) This implies that the equation

$$F_{reg}(\Gamma(z)) - F_{reg}(z) = J(z) \qquad\qquad (2.1.2)$$

with $J(z) = -2 - F_{sing}(\Gamma(z)) + F_{sing}(z) = 0(z^4)$ has a (unique) solution of the form

$$F_{reg}(z) = - \sum_{j=0}^{\infty} J(\Gamma^n(z)) \qquad\qquad (2.1.3)$$

analytic on sectors $s^{\pm}$ respectively.

ii) The representation of F_{reg} by (2.1.3) has a serious draw-
back, namely only power law convergence (we shall need to
construct effectively the solution in later sections). This
can be improved by some additional information about F_{reg} ,
which is very interesting in its own right. Namely, the inverse
Laplace transform of F_{reg} turns out to be a resurgent function, i.e. it is analytic on a Riemann surface over $\mathbb{C}$, with
singularities only on the lattice $2\pi i \mathbb{Z}$. Ecalle gives explicit bounds on this Laplace transform which we shall re-
produce below, and we can use the method of <u>Borel summation</u>
as described by <u>Loeffel</u> [33] to reconstruct F_{reg} from its
formal power series P_{reg} , using the bound on its Laplace
transform.

2.2. <u>Normal forms</u>

The transformations which we encounter in our work have germs
starting as $-z + O(z^2)$. We find it therefore useful to discuss
briefly the normal form of equations

$$F_p(G_p(z)) - F_p(z) = R_p(z) \qquad\qquad (2.2.1)$$

where

$$G_p(z) = e^{2\pi i/p} \cdot z + O(z^2) \ , \quad p = 1,2,3,\ldots$$

(thus the case of interest to us is $p = 2$). We assume G_p is
analytic near zero. The strategy to solve (2.2.1) is based on considering the equation

$$F_p(G_p^p(z)) - F_p(z) = \hat{R}_p(z) ,$$ (2.2.2)

where

$$\hat{R}_p(z) = \sum_{k=0}^{p-1} R_p(G_p^k(z))$$

and where G_p^ν is the ν-fold iterate of G_p. Clearly, for z near zero,

$$G_p^p(z) = z + \mathcal{O}(z^{p+1}) .$$ (2.2.3)

We want to reduce this to the form

$$g(z) = z - z^2 + \mathcal{O}(z^3) ,$$ (2.2.4)

and following Ecalle this is achieved by the change of coordinates $z \to z^{1/p}$, which leads to

$$(G_p(z^{1/p}))^p = z \cdot (1 + \mathcal{O}(z))^p$$

$$= z + \mathcal{O}(z^2) .$$ (2.2.5)

We assume the term in z^2 is not absent. Then a dilation of coordinates brings the first two terms of G_p^p into the desired form. However, generically, the next term will be $\mathcal{O}(z^{2+1/p})$ and not $\mathcal{O}(z^3)$ when $p \geq 2$. A term $\mathcal{O}(z^3)$ can be achieved by a further transformation which we describe only for the case $p = 2$, (suppressing the index $p = 2$ in the sequel).

We start with

$$G(z) = -z + O(z^2) \tag{2.2.6}$$

and get that, for some κ,

$$G(G(z)) = z - \frac{2}{\kappa} z^3 + O(z^4) . \tag{2.2.7}$$

To eliminate the $O(z^{5/2})$ as indicated we need $O(z^5)$ instead of $O(z^4)$ in (2.2.7). This is equivalent to saying that we need $O(z^3)$ instead of $O(z^2)$ in (2.2.6). Locally there is always a coordinate transformation T_A, such that the functions

$$G_O(z) = (T_A)^{-1} \circ G \circ T_A(z)$$

respectively

$$\Gamma(z) = (T_A)^{-1} \circ G \circ G \circ T_A(z)$$

have the desired properties. We choose

$$T_A(z) = \frac{z}{1+Az}$$

($\equiv (T_{-A})^{-1}(z)$), and an easy calculation shows that one has to set $A = -\frac{1}{4}G''(0)$.

In these new coordinates we are led to discuss the theory of equations of the form

$$F_O(\Gamma(z)) - F_O(z) = R(z) , \tag{2.2.8}$$

where, as in the introduction, the function Γ is of the form
$\Gamma(z) = z - \frac{2}{\kappa} z^3 + O(z^5)$ and we suppose that it is analytic on
a disk $D_\Gamma = D(0,\rho_\Gamma)$. Furthermore, we assume that
$\Gamma(z) = G_0(G_0(z))$, with $G_0(z) = -z + O(z^3)$ and with G_0 analytic on a disk $D_0 = D(0,\rho_0)$, $D_0 \supset D_\Gamma$. The function R in
(2.2.8) should be analytic in a domain D_R, to be described in
detail below (one should think of D_R typically as a sector with
apex at the origin).

We change variables as described above, i.e. we define

$$g(z) = \frac{4}{\kappa}(\Gamma((\tfrac{\kappa}{4}z)^{1/2}))^2 = z - z^2 + O(z^3) ,$$

$$f_0(z) = F_0((\tfrac{\kappa}{4}z)^{1/2}) ,$$

$$r(z) = R((\tfrac{\kappa}{4}z)^{1/2}) .$$

These functions are analytic in domains D_g, D_f and D_r respectively. We have for example

$$D_g = D_2(0,\tfrac{4}{\kappa}\cdot\rho_\Gamma^2) = \{z\in\mathbb{C}_2 \big| |z| < \tfrac{4}{\kappa}\rho_\Gamma^2\} ,$$

where $\mathbb{C}_2$ is the Riemann surface of $z^{1/2}$ over $\mathbb{C}$.

It may be helpful to explain in more detail our conventions.
The functions we consider below are "analytic functions of $z^{1/2}$
or $t^{1/2}$". An analytic function of $z^{1/2}$ has a convergent expansion in powers of $z^{1/2}$, near $z = 0$, and is therefore de-

fined and analytic on a domain in the two sheeted Riemann surface $\mathbb{C}_2$. We generally consider domains in which every $z \in \mathbb{C}$, $z \neq 0$ has two representations. When we talk about sectors with apex at $z = 0$, we mean the two-sheeted sectors, and the boundary is also two-sheeted. Nevertheless, we shall always write $f(z)$, etc., for analytic functions of $z^{1/2}$. Recall also, that $D_2(0,\rho)$ denotes a disk of radius ρ in the ramified plane $\mathbb{C}_2$. (An extensive discussion of such questions can be found in [13].) After these precautions, we proceed with our discussion of (2.2.8).

Using the definitions of f_o, g and r we get for (2.2.8)

$$f_o(g(z)) - f_o(z) = r(z) . \qquad (2.2.9)$$

It is this equation which we discuss now in detail. We write

$$g(z) = z - z^2 + \hat{g}(z^{1/2}) ,$$

where

$$\hat{g}(\zeta) = \sum_{k=6}^{\infty} g_k \zeta^k .$$

In the cases of interest to us, the function r will always be of the form

$$r(z) = r_o + r_2 z + r_3 z^{3/2} + \hat{r}(z^{1/2}) ,$$

where

$$\hat{r}(\zeta) = \sum_{k=4}^{\infty} r_k \zeta^k .$$

We then set

$$f_{sing}(z) = \frac{a_o}{z} + c_o \log z + d_o z^{1/2} ,$$

with

$$a_o = r_o ,$$

$$c_o = -r_2 - a_o(1 - g_6) ,$$

$$d_o = -2 \cdot (r_3 + a_o g_7) .$$

A straightforward substitution shows that

$$f_{sing}(g(z)) - f_{sing}(z) = r_o + r_2 z + r_3 z^{3/2} + j(z) ,$$

with $j(z) = \hat{j}(z^{1/2})$, and $\hat{j}(\zeta) = 0(\zeta^4)$. Therefore, writing $f_o = f_{sing} + f$, we are led to study the equation

$$f(g(z)) - f(z) = -j(z) .$$

Remark : If g has no zero on the domain of analyticity of r (and is defined there), then j is analytic on the same domain (and is bounded there if the domain is bounded).

2.3. Existence and uniqueness of sectorial solutions

A formal solution of the equation

$$f(g(z)) - f(z) = -j(z) , \qquad (2.3.1)$$

where $g(z) = z - z^2 + O(z^3)$ and $j(z) = O(z^2)$ is given by

$$f(z) = \sum_{n=0}^{\infty} j(g^n(z)) . \qquad (2.3.2)$$

We shall study the convergence of this sum. We assume $g(z)$ is an analytic function of $z^{1/2}$ which is defined on the domain D_g in the Riemann surface $\mathbb{C}_2$ of $z^{1/2}$.

Definition 2.3.1. (The set U_g of converging points.) We choose $\rho > 0$ sufficiently small so that the disk

$$D_2(0,\rho) \equiv \{z \in \mathbb{C}_2 \big| |z| < \rho\}$$

satisfies

1) $D_2(0,\rho) \subset D_g$,

2) g admits on $D_2(0,\rho)$ no fixed points and zeros other than
     $z = 0$.

Then we define U_g as the interior of the set of all $z \in D_2(0,\rho)$ for which

3) $g^n(z) \in D_2(0,\rho)$ for all $n \in \mathbb{N}$.

4) $\lim_{n \to \infty} g^n(z) = 0$.

5) $\lim_{n \to \infty} \arg(g^n(z)) = 0$.

Theorem 2.3.1. (Ecalle) If j is defined on U_g and is analytic in $z^{1/2}$ and if $j(z) = O(z^2)$ uniformly as $z \to 0$ in U_g, then the right-hand side of (2.3.2) converges uniformly on U_g and defines an analytic function f which solves (2.3.1).

To make this exposition self-contained, we give the proof
of this theorem.

Proof : The idea is very simple. Every point $z \in U_g$ is eventually
attracted to $z = 0$ with a definite "speed"

$$|g^n(z)| \le \frac{K(z)}{n} .$$ (2.3.3)

Since $j(z) = 0(z^2)$, the sum (2.3.2) converges absolutely, and
defines an analytic function f which solves (2.3.1) on U_g. We
start to show the bound (2.3.3) in certain small sectorial regions
$U(\beta)$, and argue that every point in U_g has an orbit which
eventually enters one of these sectors.

Given $\beta \in (0, \pi/2)$, there is a $\rho(\beta) \in (0, \rho)$ such that for
$z \in D_2(0, \rho(\beta))$ one has the bound

$$|\gamma_1(z) - 1| \le \sin \beta ,$$ (2.3.4)

where

$$\gamma_1(z) = \frac{1}{g(z)} - \frac{1}{z} = 1 + 0(z) ,$$ (2.3.5)

since $g(z) = z - z^2 + 0(z^3)$. (This is the place where we need
that g is not only $z - z^2 + 0(z^{5/2})$, which in turn would imply
a function γ_1 of the form $1 + 0(z^{1/2})$. γ_1 will play a crucial
role below.) We now define the two-sheeted domain

$$U(\beta) = \{z \in \mathbb{C}_2 \mid z \ne 0, \left|\arg(\frac{1}{z} - \frac{1}{r})\right| < \pi - \beta\} ,$$

where $r = r(\beta) > 0$ is chosen sufficiently small, so that

$U(\beta) \subset D_2(0,\rho(\beta))$.

Lemma 2.3.2. We have the inclusion $U(\beta) \subset U_g$ for all $\beta \in (0,\pi/2)$.

Proof : With $\gamma_1(z)$ as defined in (2.3.5), it is easy to see that

for $z \in D_2(0,\rho(\beta))$ one has from (2.3.4) :

$$|\arg(\gamma_1(z))| \leq \beta ,$$ (2.3.6)

and

$$\operatorname{Re}\gamma_1(z) \geq 1 - \sin \beta .$$ (2.3.7)

Combining the definition of $U(\beta)$ with (2.3.6), we get

$$\left|\arg\left(\frac{1}{g(z)} - \frac{1}{r}\right)\right| = \left|\arg\left(\gamma_1(z) + \frac{1}{z} - \frac{1}{r}\right)\right| < \pi - \beta ,$$

which proves $g(U(\beta)) \subset U(\beta)$, and hence $g^n(U(\beta)) \subset U(\beta)$ for all $n \in \mathbb{N}$.   Therefore we find that

$$\sum_{m=0}^{n} \gamma_1(g^m(z)) = \frac{1}{g^{n+1}(z)} - \frac{1}{z} ,$$ (2.3.8)

is defined and hence (2.3.7) implies

$$\operatorname{Re}\left(\frac{1}{g^{n+1}(z)}\right) \geq (n+1)(1-\sin \beta) + \operatorname{Re}\left(\frac{1}{z}\right) .$$ (2.3.9)

This proves that $g^n(z) \to 0$ for all $z \in U(\beta)$. Since $g(z) = z - z^2 + O(z^3)$ it follows that $\gamma_1(z) = 1 + z + O(z^2)$, so that $\gamma_1(g^n(z)) = 1 + O(\frac{1}{n})$. Thus by (2.3.8)

$$g^{n+1}(z) = \frac{1}{n+0\,(\log n + \frac{1}{|z|})} \quad ,$$

and hence $\arg(g^{n+1}(z)) \to 0$, as $n \to \infty$, and hence $U(\beta) \subset U_g$.

Lemma 2.3.2 is proved.

We continue the proof of Theorem 2.3.1. It is easy to see (cf. Fig. 2.1) that the $U(\beta)$ contain sectors of

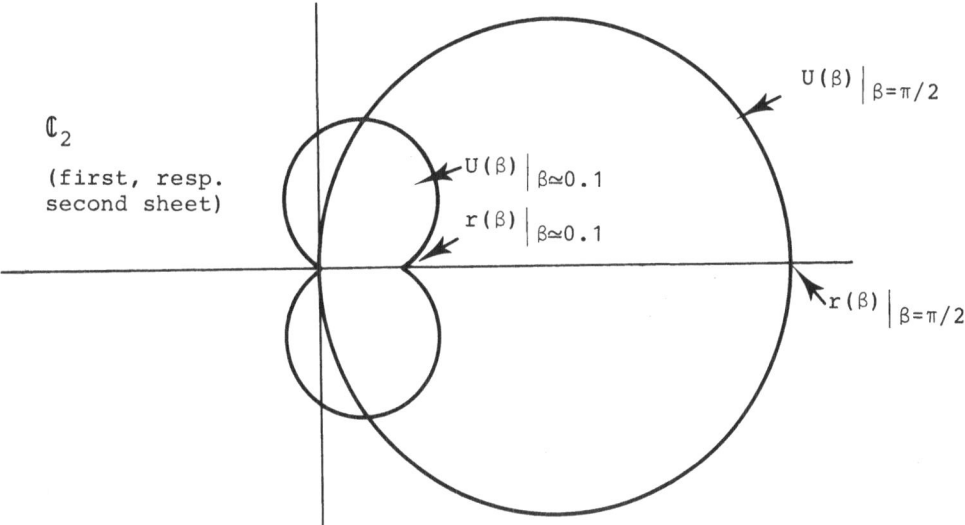

Fig. 2.1. Shape of $U(\beta)$ for $\beta = \pi/2$ and $\beta \sim \pi/10$.

half-opening angle less than $\pi - \beta$ around the real axis. Therefore if $z \in U_g$, we must have $g^n(z) \in U(\beta)$ for some β and some n. Together with the bound (2.3.9) this fact implies the assertion of the theorem. (The analyticity follows from uniform convergence on compact sets of U_g.)

2.4. Construction of sectorial solutions by Borel resummation

The sum (2.3.2) defining the solution of (2.3.1) could in

principle be used as a constructive tool for producing approximate
solutions to any degree of accuracy in a finite number of arithmetic steps. The convergence of the sum is however very slow and
we show therefore in this section that the solution (2.3.2) can
also be constructed by Borel summation (inverse Laplace transform), starting from the (unique) formal powers series solution
of (2.3.1). We recall at this point, adapted to our needs $(\mathbb{C}_2$
ramification point), the theorem of Watson and Nevanlinna on Borel
summability [42,50].

Theorem 2.4.1. Let f be analytic in the disk
$D_{2,R} = \{z \in \mathbb{C}_2 | \mathrm{Re}(z^{-1}) > R^{-1}\}$ and satisfy there the asymptotic expansion

$$f(z) = \sum_{k=1}^{n-1} a_k z^{(k+1)/2} + R_n(z) , \qquad (2.4.1)$$

with the estimates

$$|a_{n-1}| \leq L \cdot \sigma^{(n-1)/2} (\tfrac{n}{2} - 1)!^{*)} ,$$

$$(2.4.2)$$

$$|R_n(z)| \leq L \cdot \sigma^{n/2} (\tfrac{n-1}{2})! |z|^{(n+1)/2} ,$$

uniformly in n and $z \in D_{2,R}$. Then the series

---

*) To avoid confusion with the function Γ introduced above we
write $x! = \int_0^\infty t^x e^{-t} dt$ (instead of $\Gamma(x+1)$) for the Euler
gamma-function.

$$f_B(t) = \sum_{k=1}^{\infty} \frac{a_k}{(\frac{k-1}{2})!} t^{(k-1)/2} \qquad (2.4.3)$$

converges for $|t| < 1/\sigma$ and has an analytic continuation to the striplike region $S_{2,\sigma} = \{t \in \mathbb{C}_2 | \text{dist}(t, \mathbb{R}_+) < \frac{1}{\sigma}\}$. It is represented there by the contour integral

$$f_B(t) = \frac{1}{4\pi i} \oint_C e^{t/z} f(z) \frac{dz}{z^2} , \qquad (2.4.4)$$

$$c : \text{Re}(\frac{1}{z}) = \frac{1}{r} \quad (\text{in } \mathbb{C}_2)$$

where $r < R$. Furthermore f_B satisfies the bound

$$|f_B(t)| \leq K \cdot \exp(|t|/R) , \qquad (2.4.5)$$

uniformly in every $S_{2,\sigma'}$, $\sigma' > \sigma$, and the function f can be represented by the absolutely convergent integral

$$f(z) = \int_0^{\infty} e^{-t/z} f_B(t) dt , \qquad (2.4.6)$$

for any $z \in D_{2,R}$. Conversely, if $f_B(t)$ is a function analytic in $S_{2,\sigma''}$, $\sigma'' < \sigma$, and satisfying (2.4.5) on $S_{2,\sigma''}$, then the function $f(z)$ defined by (2.4.6) is analytic in $C_{2,R}$ and satisfies (2.4.1) and (2.4.2) uniformly in any $C_{2,R'}$, with $R' < R$.

Remark : As a consequence of the theorem, at most one function f which is analytic in $D_{2,R}$ can satisfy (2.4.1) and (2.4.2), for given a_k.

In the next two lemmas we state some properties of the Borel transform.

Lemma 2.4.2. Let the function f satisfy conditions (2.4.1) and (2.4.2) of Theorem 2.4.1 for some domain $D_{2,R}$. Let g be a function, analytic in $D_2(0,\rho) \supset D_{2,R}$. Suppose furthermore that $g(0) = 0$, $g(D_{2,R}) \subset D_{2,R}$, then the composed function $h = f \circ g$ is analytic in $D_{2,R}$ and satisfies again conditions (2.4.1) and (2.4.2) of Theorem 2.4.1.

Proof : It is clear that the composed function $h = f \circ g$ is analytic in $D_{2,R}$. Using furthermore that for $z \in D_2(0,\rho)$ the function g can be bounded by $|g(z)| < L|z|$ for some constant L, conditions (2.4.1) and (2.4.2) of Theorem 2.4.1 can be verified using the Cauchy formula.

Lemma 2.4.3. Let $f_{1,B}$ and $f_{2,B}$ be two Borel functions, i.e. satisfy conditions (2.4.5) of Theorem 2.4.1, then the composition law $\bullet$, which we define by the equation

$$(f_{1,B} \bullet f_{2,B})(t) = f_{1,B}(t) + \sum_{n \geq 1} \frac{1}{n!}(f_{2,B}(t) + \delta(t))^{*n}$$

$$* ((-t)^n f_{1,B}(t))$$

is defined and the function $f_{1,B} \bullet f_{2,B}$ is again a Borel function.

Proof : Left to the reader. (See [13] for more details about $\bullet$.)

Remark : Let f_1, f_2 and h be the functions which are associated by (2.4.6) to $f_{1,B}$, $f_{2,B}$ and $f_{1,B} \cdot f_{2,B}$ respectively, then

<u>formally</u> $h(z) = f_1(z/(1 + z \cdot (1+f_2(z))))$.

After these preparatory steps we are now ready to state and prove the main theorem of this section. We assume

1) g is analytic in $z^{1/2}$ near $z = 0$ and maps U_g into itself as in Theorem 2.3.1,

2) j is analytic in U_g and satisfies bounds of the form (2.4.1) and (2.4.2) in this domain. Furthermore $j(z) = O(z^2)$ as $z \to 0$ in U_g.

Then we have the following theorem.

<u>Theorem 2.4.4.</u> <u>Under the above assumptions, the solution f of the equation</u>

$$f(g(z)) - f(z) = -j(z)$$

<u>can be represented for every</u> $z \in D_{2,R} \subset U_g$ <u>as</u>

$$f(z) = \int_0^\infty e^{-t/z} f_B(t) \, dt \ ,$$

<u>where</u> f_B <u>can be explicitly calculated from the Taylor coefficients of g and j. R depends on g, j and U_g. In particular f is Borel summable.</u>

<u>Proof</u> : We proceed as follows. First we deduce a formal power series expression $\hat{f}_B$ for the function f_B. Then we show that this expression defines a function which satisfies all the conditions of

Theorem 2.4.1 to be the Borel transform of a function f.   Then

we show that the function f so constructed solves Ecalle's

equation, at least on domains $D_{2,R}$, R sufficiently small. To

start this program we denote

$$\mathfrak{j}(x) \;=\; \sum_{k=3}^{\infty} j_k x^{(k+1)/2}$$

the asymptotic expansion of the function j, and

$$\mathfrak{g}(x) \;=\; x \,-\, x^2 \,+\, x^3 \cdot \sum_{k=1}^{\infty} g_k x^{(k-1)/2}$$

the expansion of g at z = 0.   It is now easy to see that the

formal power series equation

$$\hat{f}(\hat{g}(x)) \,-\, \hat{f}(x) \;=\; -\hat{j}(x) \tag{2.4.7}$$

has exactly one formal power series solution of the form

$$\hat{f}(x) \;=\; \sum_{k=1}^{\infty} f_k x^{(k+1)/2}$$

In a next step we introduce the formal power series L^+ and L^-,

which we define by the equation

$$L^{\pm}(x) \;=\; \frac{x}{1 \pm x} \; .$$

We have the relations $L^{\pm}(L^{\mp}(x)) = x$.   We now define the formal

power series $\hat{g}_0$ by the equation

$$\hat{g}_o(x) = L^-(\hat{g}(x)) = \frac{\hat{g}(x)}{1-\hat{g}(x)} = x + O(x^3) \ .$$

(In this way we have taken advantage of the fact that, expressed in $\hat{z} = \frac{1}{z}$, the function g is almost a "translation", i.e. $1/g(\hat{z}) = \hat{z} + 1 + O(\frac{1}{\hat{z}})$.) Evidently, $\hat{g}_o$ is the expansion of the function $g_o(z) = g(z)/(1-g(z))$ at $z = 0$. Reexpressing (2.4.7) in terms of $\hat{g}_o$ we get

$$\hat{f}(L^+(\hat{g}_o(x))) - \hat{f}(x) = -\hat{j}(x) \ . \tag{2.4.8}$$

Next we define operators $\mathbb{L}^{\pm}$ and Γ_o. Let φ be a formal power series of the type $\varphi(x) = \sum_{k\geq 1} \varphi_k x^{(k+1)/2}$, then we define

$$(\mathbb{L}^{\pm}\varphi)(x) = \varphi(L^{\pm}(x)) \ ,$$

$$(\Gamma_o\varphi)(x) = \varphi(\hat{g}_o(x)) \ .$$

In terms of those operators (2.4.8) is equivalent to

$$\Gamma_o \mathbb{L}^+\hat{f}(x) = \hat{f}(x) - \hat{j}(x) \ .$$

Manipulating this equation formally, we get successively

$$(\Gamma_o-1)\mathbb{L}^+\hat{f}(x) = (1-\mathbb{L}^+)\hat{f}(x) - \hat{j}(x) \ ,$$

$$(\Gamma_o-1)\mathbb{L}^+(1-\mathbb{L}^+)^{-1}\hat{f}(x) = (\Gamma_o-1)(\mathbb{L}^--1)^{-1}\hat{f}(x)$$

$$= \hat{f}(x) - (1-\mathbb{L}^+)^{-1}\hat{j}(x) \ ,$$

and finally

$$\hat{f}(x) = (1-\mathbb{L}^+)^{-1}\hat{j}(x) + \sum_{n\geq 1} (1-\mathbb{L}^+)^{-1}[(\Gamma_o-1)(\mathbb{L}^--1)^{-1}]^n\hat{j}(x) .$$

$$(2.4.9)$$

We denote by $\psi(x) = (\mathbb{L}^--1)^{-1}\varphi(x)$ the unique power series solution without constant term of the equation

$$(\mathbb{L}^--1)\psi(x) = \varphi(x) ,$$

i.e. we solve the equation $\psi(\frac{x}{1-x}) - \psi(x) = \varphi(x)$. Remember that by assumption $\hat{j}(x) = 0(x^2)$. Therefore if we set

$$\varphi(x) = \mathbb{L}^-\hat{j}(x) = 0((\frac{x}{1-x})^2) = 0(x^2) ,$$

then, since $(1-\mathbb{L}^+)^{-1} = (\mathbb{L}^- - 1)^{-1}\mathbb{L}^-$, we have to solve equations of the form

$$\psi(\frac{x}{1-x}) - \psi(x) = 0(x^2) ,$$

and it is easy to see that the power series ψ can be calculated coefficient by coefficient, starting with $\psi(x) = 0(x)$. In general if $\varphi(x) = 0(x^{n/2})$, then $\psi(x) = 0(x^{n/2-1})$. Let us now concentrate on the sum

$$\psi(x) = \sum_{n\geq 1} (1-\mathbb{L}^+)^{-1}[(\Gamma_o-1)(\mathbb{L}^--1)^{-1}]^n\hat{j}(x) .$$

Remember that by construction, $\Gamma_o(x) = \hat{g}_o(x) = x + 0(x^3)$, and therefore we get if $n = 1$

$$(1-\mathbb{L}^+)^{-1}(\Gamma_o-1)(\mathbb{L}^--1)^{-1}0(x^2)$$

$$= (\Gamma_o-1)(\mathbb{L}^--1)^{-1}0(x) = (\mathbb{L}^--1)^{-1}0(x^3) = 0(x^2).$$

and more generally

$$(\Gamma_o-1)(\mathbb{L}^--1)^{-1}0(x^{n/2}) = (\mathbb{L}^--1)^{-1}0((x+0(x^3))^{n/2} - x^{n/2})$$

$$= (\mathbb{L}^--1)^{-1}0(x^{n/2+2}) = 0(x^{n/2+1}).$$

From that we conclude that

$$(1-\mathbb{L}^+)^{-1}[(\Gamma_o-1)(\mathbb{L}^--1)^{-1}]^n \hat{\jmath}(x) = 0(x^{2+n-1}) = 0(x^{n+1}),$$

and therefore the above sum ψ makes sense as a formal power series. Therefore (2.4.9) leads to the formal power series solution of (2.4.7), explicitly expressed in terms of the power series $\hat{g}$ and $\hat{\jmath}$. In a next step we want to pass to the (formal) Borel transform $\hat{f}_B$ of $\hat{f}$. We need a more explicit version of the expression (2.4.9). We define the formal power series $\hat{\gamma}$ by the equation

$$\hat{\gamma}(x) = \frac{1}{\hat{g}_o(x)} - \frac{1}{x}.$$

($\hat{\gamma}$ is the expansion of the function $\gamma(z) = \gamma_1(z) - 1$ at $z = 0$, where $\gamma_1(z)$ is defined through (2.3.5).) We have that $\hat{\gamma}(x) = (x + 0(x^3))^{-1} - \frac{1}{x} = 0(x)$. In terms of $\hat{\gamma}$ we can calculate the action of the operator (Γ_o-1) explicitly. Let

$$\psi(x) \equiv (\Gamma_0 - 1)\varphi(x) = \varphi(\hat{g}_0(x)) - \varphi(x)$$

$$= \varphi\left(\frac{x}{1+x\hat{\gamma}(x)}\right) - \varphi(x) \ .$$

Define now $\psi^\infty(x) = \psi(\frac{1}{x})$ and $\varphi^\infty(x) = \varphi(\frac{1}{x})$, then

$$\psi^\infty(x) = \varphi^\infty(x + \hat{\gamma}(\frac{1}{x})) - \varphi^\infty(x)$$

$$= \sum_{n \geq 1} \frac{1}{n!}(\hat{\gamma}(\frac{1}{x}))^n [(\frac{d}{dy})^n \varphi^\infty(y)] \Big|_{y=x}$$

and therefore we see that

$$(\Gamma_0 - 1) = \sum_{n \geq 1} \frac{1}{n!}(\hat{\gamma}(x))^n (\frac{d}{d(1/x)})^n \ .$$

We introduce the operators $\hat{\mathbb{H}}_m$:

$$(\hat{\mathbb{H}}_m \varphi)(x) = \frac{1}{m!}(\hat{\gamma}(x))^m (\frac{d}{d(1/x)})^m ((\mathbb{L}^- - 1)^{-1} \varphi(x)$$

In terms of these operators we have

$$\hat{f}(x) = \sum_{n \geq 0} (1-\mathbb{L}^+)^{-1} [(\Gamma_0 - 1)(\mathbb{L}^- - 1)^{-1}]^n$$

$$= (1-\mathbb{L}^+)^{-1}\hat{g}(x) + (1-\mathbb{L}^+)^{-1} \sum_{n \geq 1} \sum_{m_1 \geq 1} \cdots \sum_{m_n \geq 1} \hat{\mathbb{H}}_{m_1} \cdots \hat{\mathbb{H}}_{m_n} \hat{g}(x) \ .$$

$$(2.4.10)$$

We are now prepared to pass on a formal level to the Borel transform. Note that on the power series level we have that

$$\hat{f}(x) = \int_0^\infty e^{-t/x} \hat{f}_B(t)\,dt \ .$$

Similarly,

$$\mathbb{L}^{\pm}\hat{f}(x) = f(\frac{x}{1\pm x}) = \int_{o}^{\infty}e^{-t/x}(e^{\mp t}\hat{f}_B(t))dt ,$$

which means that the (formal) Borel transform of the operators $\mathbb{L}^{\pm}$ is just multiplication by $e^{\mp t}$. Taking therefore the Borel transform of (2.4.10) leads to

$$\hat{f}_B(t) = \frac{1}{1-e^{-t}}\{\hat{j}_B(t) + \sum_{\substack{n\geq 1 \\ m_1\geq 1 \\ \cdots \\ m_n\geq 1}}(\mathbb{H}_{m_1}*\cdots*\mathbb{H}_{m_n})\hat{j}_B(t)\} ,$$

where we have used the definitions

$$\mathbb{H}_m\varphi(t) \equiv (\hat{\gamma}_B)^{*m} * (E_m\varphi)(t) ,$$

$$E_m(t) \equiv \frac{(-t)^m}{m!}\frac{1}{e^t-1} .$$

Writing all convolutions explicitly we have finally

$$\hat{f}_B(t) = \frac{1}{1-e^{-t}}\{\hat{j}_B(t) + \sum_{\substack{n\geq 1 \\ m_1\geq 1 \\ \cdots \\ m_n\geq 1}}\int_o^t dt_1\cdots\int_o^{t_{n-1}}dt_n$$

$$\cdot([(\hat{\gamma})^{m_1}]_B(t-t_1) E_{m_1}(t_1)\cdots$$

$$[(\hat{\gamma})^{m_n}]_B(t_{n-1}-t_n) E_{m_n}(t_n)\hat{j}_B(t_n))\} . \tag{2.4.11}$$

We now show that the formal expression $\hat{f}_B$ defines a function f_B which satisfies conditions (2.4.1) and (2.4.2) of Theorem 2.4.1, and which therefore is the Borel transform of some function f. Remember that the function $g(z)$ is analytic in $z^{1/2}$ near $z = 0$, and can therefore be represented there by a convergent power series in $z^{1/2}$. As a consequence this is also true for the functions $g_o(z)$ and $\gamma(z)$. It follows *) that the radius of convergence of the series for the Borel transform $\gamma_B(t)$ of $\gamma(z)$ is infinite, i.e. $\gamma_B(t)$ can be represented on all of $\mathbb{C}_2$ by a convergent series in $t^{1/2}$. Furthermore one has

$$|\gamma_B(t)| \le C_1 \exp(C_2|t|) ,$$

$C_1 > 0$, $C_2 > 0$ and $t \in \mathbb{C}_2$. Using this, we easily bound

$$|(\gamma_B)^{*m}(t)| \le C_1^m \cdot \exp(C_2|t|) \frac{|t|^{m-1}}{(m-1)!} , \quad t \in \mathbb{C}_2 .$$

We also give bounds on the factors $E_m(t)$ in (2.4.11), but we need two variants. Given any finite are C in $\mathbb{C}_2 \backslash 2\pi i \mathbb{Z}$, one easily sees that there are constants $C_3(C)$, $C_4(C)$ such that

$$|E_m(t)| \le C_3 C_4^m .$$

(Better bounds will be given later.) On the other hand, on any infinite ray J_θ of the form $\text{Re } t > 0$, $\arg t = \theta$, $|\theta| < \pi/2$ we have the bound

---

*)
This is the place where we need that $g(z)$ is of the form $g(z) = z - z^2 + O(z^3)$ (and not only $z - z^2 + O(z^{5/2})$). $\gamma(z)$ is therefore of the form $\gamma(z) = O(z)$ (and not only $O(z^{1/2})$) and $\gamma_B(t)$ is defined.

$$|E_m(t)| = |\frac{t^m}{m!}\frac{1}{e^t-1}| \leq (\frac{1}{\cos\theta})^m \frac{(|t|\cos\theta)^m/m!}{e^{|t|\cdot\cos\theta}-1} \leq (\frac{1}{\cos\theta})^m \;,$$

so that $C_3(J_\theta) = 1$, $C_4(J_\theta) = (\cos\theta)^{-1}$. (Similar bounds, ex-changing L^+ with L^- throughout could be obtained in the left half plane, see [13] for more details.)

Next we note that the assumptions on j imply that j_B exists and satisfies a bound of the form

$$|j_B(t)| \leq C_j \cdot |t| \cdot \exp(E_j \cdot |t|) \;,$$

$C_j > 0$, $E_j > 0$. We assume $E_j \geq C_2$ (otherwise make it larger). Inserting all estimates we get, using

$$e^{C_2|t-t_1|}e^{C_2|t_1-t_2|}\ldots e^{C_2|t_n|} = e^{C_2|t|}$$

(for $t \geq t_1 \geq \ldots \geq t_n \geq 0$) and $m = \sum\limits_{j=1}^{n} m_j$, the estimate

$$|\int_0^t dt_1 \ldots \int_0^{t_{n-1}} dt_n (\gamma^{m_1})_B(t-t_1)E_{m_1}(t_1)\ldots$$

$$(\gamma^{m_n})_B(t_{n-1}-t_n)E_{m_n}(t_n)j_B(t_n)|$$

$$\leq C_j e^{E_j|t|} \quad (C_1 C_4)^m C_3^n$$

$$\cdot \int_0^{|t|} dt_1 \ldots \int_0^{t_{n-1}} dt_n \frac{(|t|-t_1)^{m_1-1}}{(m_1-1)!} \frac{(t_1-t_2)^{m_2-1}}{(m_2-1)!} \ldots$$

$$\frac{(t_{n-1}-t_n)^{m_n-1}}{(m_n-1)!} \cdot t_n = C_j e^{E_j|t|} (C_1 C_4)^m C_3^n \frac{|t|^{m+1}}{(m+1)!} \;.$$

We can now bound the sum in (2.4.11) by $C_j e^{E_j |t|}$ times the following expression

$$\sum_{n \geq 1} \sum_{m_i \geq 1} (C_1 C_4)^m C_3^n \frac{|t|^{m+1}}{(m+1)!} = \sum_{n \geq 1} \sum_{m=n}^{\infty} \binom{m-1}{n-1} \frac{|t|^{m+1}}{(m+1)!} C_3^n (C_1 C_4)^m$$

$$= \sum_{m=1}^{\infty} (\sum_{n=1}^{m} \binom{m-1}{n-1} C_3^{n-1}) C_3 (C_1 C_4)^m \frac{|t|^{m+1}}{(m+1)!}$$

$$= \frac{C_3}{1+C_3} \sum_{m=1}^{\infty} (C_1 C_4 (1+C_3))^m \frac{|t|^{m+1}}{(m+1)!}$$

$$\leq \frac{C_3}{1+C_3} \cdot |t| \cdot \{ \exp(C_1 C_4 (1+C_3) |t|) - 1 \} \qquad (2.4.15)$$

and get finally

$$|\hat{f}_B(t)| \leq C_j e^{E_j |t|} C_3 C_4 e^{|t|}$$

$$+ \frac{C_3 C_j e^{E_j |t|}}{1+C_3} \cdot \{ e^{C_1 C_4 (1+C_3) |t|} - 1 \} \cdot C_3 C_4 e^{|t|} \qquad (2.4.16)$$

In other terms,

$$|\hat{f}_B(t)| \leq C_f e^{E_f \cdot |t|} , \qquad (2.4.17)$$

where

$$C_f = C_j C_3 C_4 \qquad (2.4.18)$$

$$E_f = E_j + C_1 C_4 (1+C_3) + 1 ,$$

and C_3 , C_4 depend on the path (or arc) chosen. This proves that the series $\hat{f}_B$ converges uniformly on every J_θ, and every finite arc C and defines a function f_B, which satisfies the hypothesis (2.4.5) of Theorem 2.4.1 and is therefore the Borel transform of a function f,

$$f(z) = \int_o^\infty e^{-t/z} f_B(t) \, dt , \qquad (2.4.19)$$

which is analytic on $D_{2,E_f^{-1}}$. Choose now $R < E_f^{-1}$, such that $D_{2,R} \subset U_g$ and $g(D_{2,R}) \subset D_{2,R}$. (See Section 2.3 for the properties of U_g.) The function $f \circ g$ is defined on $D_{2,R}$ and Borel summable by Lemma 2.4.2. Furthermore we have that $g(z) = z/(1+z(1+\gamma(z)))$ and therefore by construction

$$f_B \bullet \gamma_B - f_B = -j_B ,$$

and by Lemma 2.4.3 $f_B \bullet \gamma_B$ is again a Borel function, i.e. we can define

$$h(z) = \int_o^\infty e^{-t/z} f_B \bullet \gamma_B (t) \, dt .$$

Since $h(z)$ and $f \circ g(z)$ have the same asymptotic expansion it follows that $h = f \circ g$ and therefore f solves the equation $f(g) - f = -j$.

This completes the proof of Theorem 2.4.4.

2.5. Some properties of the sectorial solutions

In this section we complete the discussion of equations (2.1.1), (2.2.1), (2.2.8) and (2.2.9), respectively. The solution of (2.2.9) is given by the function f_o ,

$$f_o(z) = \frac{a_o}{z} + c_o \log z + d_o z^{1/2} + f(z) ,$$

where the function f can be expressed either by (2.3.2) for $z \in U_g$, or by (2.4.19) for $z \in D_{2,R} \subset U_g$. Define now the two domains

$$D_{\pm,R} = \{z \in \mathbb{C} | \operatorname{Re} z \gtrless 0, \ z^2 \in D_{2,R}\}$$

then the function $F_o(z) = f_o(\frac{4}{\kappa}z^2)$ solves (2.2.8) on $D_{\pm,R}$. To be more explicit about the function F_o , we use for the Borel transform f_B of f the decomposition

$$f_B(t) = f_{1,B}(t) + \frac{1}{t^{1/2}} f_{2,B}(t) ,$$

and we get, for $z \in D_{\pm,R}$

$$F_o(z) = \frac{\kappa a_o/4}{z^2} + c_o \log (\frac{4}{\kappa}z^2) + d_o \sqrt{\frac{4}{\kappa}} z$$

$$+ z^2 \cdot F_1(z^2) + z^3 \cdot F_2(z^2) ,$$

where the functions F_1 ,

$$F_1(z) = \frac{4/\kappa}{z} \int_o^\infty e^{-t/z} f_{1,B}(\frac{4}{\kappa}t) dt ,$$

and F_2 ,

$$F_2(z) = \frac{(4/\kappa)^{1/2}}{z} \int_o^\infty e^{-t/z} f_{2,B}(\tfrac{4}{\kappa}t) \cdot \frac{dt}{(zt)^{1/2}}$$

are analytic on $(D_R)^2 = \{z^2 | z \in D_{+,R}\} \equiv \{z^2 | z \in D_{-,R}\}$.

We discuss now (2.2.1) in the case $p = 2$. (We need this result for our discussion of the case $N < \infty$.) We have already shown that $F_o(G_o(G_o(z))) - F_o(z) = R(z)$ for $z \in D_{\pm,R}$. In addition to the assumptions we have made we suppose now $R(z) = R_o(z) + R_o(G_o(z))$ for $z \in D_{\pm,R}$. Then we define the function

$$\tilde{F}_o(z) = F_o(G_o(z)) - R_o(z) \ .$$

It is easy to check that not only $G_o(G_o(D_{\pm,R})) \subset D_{\pm,R}$, but even $G_o(D_{\pm,R}) \subset D_{\mp,R}$, and therefore $\tilde{F}_o$ is analytic on $D_{\pm,R}$. Furthermore

$$\tilde{F}_o(G_o(G_o(z))) - \tilde{F}_o(z) =$$

$$= (F_o(G_o^3(z)) - R_o(G_o^2(z))) - (F_o(G_o(z)) - R_o(z))$$

$$= ([F_o(G_o(z)) + R_o(G_o(z)) + R_o(G_o^2(z))] - R_o(G_o^2(z)))$$

$$- F_o(G_o(z)) + R_o(z) = R_o(z) + R_o(G_o(z)) = R(z) \ ,$$

and therefore, since the solution (2.2.8) is unique (up to a trivial additive constant) and using the definition of $\tilde{F}_o$, we get

$$F_o(z) = \widetilde{F}_o(z) + c = F_o(G_o(z)) - R_o(z) + c \; .$$

Iterating this equality once we get

$$F_o(z) = (F_o(G_o(G_o(z))) - R_o(G_o(z)) + c) - R_o(z) + c$$

$$= F_o(G_o(G_o(z))) - R(z) + 2c \; ,$$

and it follows that $c = 0$ and therefore

$$F_o(G_o(z)) - F_o(z) = R_o(z) \; ,$$

i.e. F_o solves (2.2.1) on $D_{\pm,R}$ and in particular (2.1.1) if we put $R_o(z) \equiv -1$.

This completes our discussion of Ecalle's work.

3. The constructive method of Loeffel

This chapter may be read independently of the remainder of the book. We explain here Loeffel's method for evaluating Laplace integrals, but adapt it for our needs in later chapters.

In the course of considering Borel summability one is confronted with the problem of evaluating constructively integrals of the form

$$f(z) = \int_0^\infty e^{-t/z} f_B(t) \, dt \; . \tag{3.1}$$

Here, f_B is a function which is analytic in a domain containing $t = 0$ and a strip of uniform width around the positive real axis, and it satisfies in this domain a bound of the type

$$|f_B(t)| \le C \cdot \exp\left(\frac{|t|}{R}\right) \; .$$

It follows that the function f is analytic in the disk $D_R = \{z \in \mathbb{C} \mid \mathrm{Re}\left(\frac{1}{z}\right) > \frac{1}{R}\}$. Loeffel [33] noted that (3.1) can be efficiently computed from the Taylor series of f_B at $t = 0$, by using a conformal mapping. In particular, if z_0 is in D_R then the Taylor coefficients of f at z_0 can be computed using this method. A special case of this method was found earlier by Nörlung, see e.g. Doetsch [43] ("Fakultätenreihen"). Here we adapt it to the case of functions f_B which are analytic as functions of $t^{1/2}$.

We suppose that we are given a function $f_B(t)$ which is analytic near $t = 0$ in $\mathbb{C}_2$, the Riemann surface of $t^{1/2}$. Furthermore, we suppose that this function f_B has an analytic continuation in strips around the positive real axis on both sheets, and we denote $f_{B,\pm}$ the two determinations of f_B. Then it is useful to write

$$f_{B,\pm}(t) = f_{B,1}(t) \pm \frac{1}{t^{1/2}} f_{B,2}(t) \ , \tag{3.2}$$

and it follows, since

$$f_{B,2}(t) = \frac{1}{2}(f_{B,+}(t) + f_{B,-}(t))$$

and

$$f_{B,2}(t) = \frac{t^{1/2}}{2} \cdot (f_{B,+}(t) - f_{B,-}(t))$$

that $f_{B,1}$ and $f_{B,2}$ are analytic in strips in $\mathbb{C}$ and that $f_{B,1}(t) = O(f_{B,\pm}(t))$, $f_{B,2}(t) = O(t \cdot f_{B,\pm}(t))$ near $t = 0$. Furthermore, if

$$|f_{B,\pm}(t)| \leq C \cdot \exp\left(\frac{|t|}{R}\right) \ ,$$

then, using the definition of $f_{B,1}$, $f_{B,2}$, we get the bounds

$$|f_{B,1}(t)| \leq C \cdot \exp\left(\frac{|t|}{R}\right) \ , \tag{3.3}$$

and

$$|f_{B,2}(t)| \leq C \cdot |t|^{1/2} \cdot \exp\left(\frac{|t|}{R}\right) \ . \tag{3.4}$$

We define now functions $f_\pm$ by the equation

$$f_{\pm}(z) = \int_0^{\infty} e^{-t/z} f_{B,\pm}(t) \, dt .$$

Using (3.2) and the bounds (3.3), (3.4), we find

$$f_{\pm}(z) = \int_0^{\infty} e^{-t/z} (f_{B,1}(t) \pm \frac{1}{t^{1/2}} f_{B,2}(t)) \, dt = f_1(z) \pm f_2(z) ,$$

where

$$f_1(z) = \int_0^{\infty} e^{-t/z} f_{B,1}(t) \, dt , \qquad\qquad (3.5)$$

and

$$f_2(z) = \int_0^{\infty} e^{-t/z} f_{B,2}(t) \frac{dt}{t^{1/2}} \qquad\qquad (3.6)$$

We first explain Loeffel's method for (3.5), which is the "standard case" and then show how the additional complication due to the factor $t^{-1/2}$ in (3.6) can be treated.

We define the conformal mapping

$$\varphi_{\beta}(\tau) = -\beta \cdot \log(1-\tau) , \quad \beta > 0 ,$$

which maps the unit disk onto a strip S_{β} ,

$$S_{\beta} = \{t = \varphi_{\beta}(\tau) \,|\, |\tau| < 1\} ,$$

and the segment $[0,1]$ onto $[0,\infty]$, with $\varphi_{\beta}(0) = 0$. The inverse map φ_{β}^{-1} is

$$\varphi_{\beta}^{-1}(\tau) = 1 - e^{-\tau/\beta} .$$

We choose now β sufficiently small, so that S_{β} is in the domain of analyticity of $f_{B,1}$. We denote by

$$f_{B,1}(t) = \sum_{k=0}^{\infty} f_{B,1}^{(k)} t^k$$

the Taylor expansion of $f_{B,1}$ at $t = 0$. Then the rearranged function

$$\hat{f}_{B,1}(t) = f_{B,1}(\varphi_\beta(t))$$

is analytic in the unit disk and the coefficients $\hat{f}_{B,1}^{(k)}$ of the Taylor expansion

$$\hat{f}_{B,1}(t) = \sum_{k=0}^{\infty} \hat{f}_{B,1}^{(k)} t^k$$

can be calculated as linear combinations of the coefficients $f_{B,1}^{(1)}, \ldots, f_{B,1}^{(k)}$. The function f_1 can now be represented as

$$f_1(z) = \int_0^\infty e^{-t/z} f_{B,1}(t) \, dt = \int_0^\infty e^{-t/z} \hat{f}_{B,1}(\varphi_\beta^{-1}(t)) \, dt =$$

$$= \int_0^\infty e^{-t/z} \sum_{k=0}^{\infty} \hat{f}_{B,1}^{(k)} (\varphi_\beta^{-1}(t))^k \, dt .$$

Using the Lebesgue dominated convergence theorem to exchange integration and summation, we get

$$f_1(z) = \sum_{k=0}^{\infty} \hat{f}_{B,1}^{(k)} m_{1,k}(z) , \tag{3.7}$$

where

$$m_{1,k}(z) = \beta \int_0^\infty e^{-\beta t/z} (1 - e^{-t})^k \, dt = \beta \cdot \frac{k!}{\prod\limits_{j=0}^{k} (\frac{\beta}{z}+j)} . \tag{3.8}$$

Since $f_{B,1}$ is bounded by (3.3), we get the bound :

$$\hat{f}_{B,1}^{(k)} = \frac{1}{2\pi i} \oint_{|t|=\rho<1} f_{B,1} (-\beta \log (1-t)) \frac{dt}{t^{k+1}} ,$$

and hence

$$|\hat{f}_{B,1}^{(k)}| \le C \cdot \exp(-\frac{\beta}{R} \cdot \log(1-\rho) - k \cdot \log \rho) = \frac{C}{(1-\rho)^{\beta/R} \cdot \rho^k}$$

and minimizing with respect to ρ we get

$$|\hat{f}_{B,1}^{(k)}| \le C \cdot k^{\beta/R} \cdot (\frac{1}{k} + \frac{1}{\beta/R})^{\beta/R} \cdot (1 + \frac{\beta/R}{k})^k . \qquad (3.9)$$

Next we bound $m_{1,k}$, by using Stirling's equation. We get, using the notation $\sigma = \beta/Rez$,

$$|m_{1,k}(z)| \le (\frac{2\pi}{1/k+1/\sigma})^{1/2} \frac{(1+\frac{\sigma}{k})^{-k} (1+\frac{1}{12\sigma-1}) (1+\frac{1}{12k-1})}{k^\sigma (\frac{1}{k} + \frac{1}{\sigma})} . \qquad (3.10)$$

The above estimates allow us to bound $\sum\limits_{k=M+1}^{\infty} \hat{f}_{B,1}^{(k)} m_{1,k}(z)$ by replacing the sum by an integral and we obtain

$$\left| \sum_{k=M+1}^{\infty} \hat{f}_{B,1}^{(k)} m_{1,k}(z) \right| \le C \cdot (\frac{1}{(\beta/R)} + \frac{1}{M})^{\beta/R} \cdot e^{\beta/R}$$

$$\cdot (1 + (1+\frac{1}{2M})^{\beta/R}) \frac{\beta}{\sigma} \sqrt{2\pi} \sqrt{\frac{1}{\beta/R}} \frac{1}{(1+\frac{\sigma}{M})^M} \cdot$$

$$\cdot \sigma^\sigma (1+\frac{1}{12\sigma-1}) (1+\frac{1}{12M-1}) \frac{1}{\beta(\frac{1}{Rez}-\frac{1}{R})-1} \cdot$$

$$\cdot M^{-\beta(\frac{1}{Rez} - \frac{1}{R})+1} \qquad (3.11)$$

and (3.8) converges for all z in the disk $D_{R'} = \{z \in \mathbb{C} \mid \mathrm{Re}\,(\frac{1}{z}) > \frac{1}{R'}\}$ with $R' = (R^{-1} + \beta^{-1})^{-1}$. Note that the functions $m_{1,k}$ are quite simple (inverse factorials), and that for $z_0 \in D_{R'}$ we may expand $m_{1,k}$ in a Taylor series around z_0, i.e.

$$m_{1,k}(z_0 + z) = \sum_{\ell=0}^{\infty} m_{1,k}^{(\ell)} z^{\ell} .$$

Furthermore, we can expand the function f_1 in a Taylor series around z_0,

$$f_1(z_0 + z) = \sum_{k=0}^{\infty} \hat{f}_{B,1}^{(k)} \left(\sum_{\ell=0}^{\infty} m_{1,k}^{(\ell)} z^{\ell} \right) = \sum_{\ell=0}^{\infty} \sum_{k=0}^{\infty} \hat{f}_{B,1}^{(k)} m_{1,k}^{(\ell)} z^{\ell} ,$$

and the second equality is true since the sum (3.7) is absolutely convergent.

Note that the functions $m_{1,k}$ depend only on the choice of the conformal mapping and not on f. For different choices of the conformal map one gets different functions and different convergence properties in (3.7). All this is described in detail in [33].

We now adapt the method to the case of (3.6) where the above procedure is slightly complicated by the presence of the factor $t^{-1/2}$ in the kernel of the integral. (We thank H. Epstein for some illuminating remarks concerning this problem.) If we would proceed exactly as before i.e. rearrange $f_{1,B}$ by exactly the same conformal mapping as above we would be led to evaluate instead of (3.8) the integrals

$$m_{2,k}(z) = \beta^{1/2} \cdot \int_o^\infty e^{-\beta t/z} (1 - e^{-t})^k \frac{dt}{t^{1/2}}$$

for $k = 1, \dots$. Although this is possible in principle it is somewhat unconvenient in practice and we therefore choose a different procedure. We define

$$\tilde{f}_{B,2}(t) = \frac{1}{\pi^{1/2}} \cdot \int_o^\infty \frac{dx}{(1+x)^2 \sqrt{x}} \, f'_{B,2} \left(\frac{t}{1+x}\right) , \qquad (3.12)$$

and if we define the $f_{B,2}^{(k)}$ by

$$f_{B,2}(t) = \sum_{k=1}^\infty f_{B,2}^{(k)} t^k ,$$

then we find

$$\tilde{f}_{B,2}(t) = \sum_{k=1}^\infty \tilde{f}_{B,2}^{(k)} t^{k-1} ,$$

where

$$\tilde{f}_{B,2}^{(k)} = \frac{(k-1/2)!}{(k-1)!} \, f_{B,2}^{(k)} \qquad \qquad {}^{*)}$$

The transformation (3.12) has an inverse, namely

$$f_{B,2}(t) = \frac{t}{\pi^{1/2}} \int_o^\infty \frac{dx}{(1+x)^{3/2} \sqrt{x}} \, \tilde{f}_{B,2}\left(\frac{t}{1+x}\right)$$

and we can calculate f_2 in terms of $\tilde{f}_{B,2}$ instead of $f_{B,2}$ as follows :

${}^{*)}$ We use the notation $x! = \int_o^\infty t^x e^{-t} dt$ (instead of $\Gamma(x+1)$) for the Euler gamma-function.

$$f_2(z) = \int_0^\infty e^{-t/z} \hat{f}_{B,2}(t) \frac{dt}{t^{1/2}}$$

$$= \int_0^\infty e^{-t/z} (\frac{1}{\pi^{1/2}} \int_0^\infty \frac{dx}{(1+x)^{3/2}\sqrt{x}} \tilde{f}_{B,2}(\frac{t}{1+x})) t^{1/2} dt$$

$$= \int_0^\infty e^{-t/z} \tilde{f}_{B,2}(t) \frac{1}{\pi^{1/2}} t^{1/2} dt \int_0^\infty e^{-tx/z} \frac{dx}{\sqrt{x}}$$

$$= z^{1/2} \int_0^\infty e^{-t/z} \tilde{f}_{B,2}(t) dt \ . \tag{3.13}$$

We proceed now exactly as in the case of f_1 , i.e. we define

$$\hat{f}_{B,2}(t) = \tilde{f}_{B,2}(\varphi_\beta(t)) = \sum_{k=0}^\infty \hat{f}_{B,2}^{(k)} t^k \ ,$$

and again the Taylor coefficients of $\hat{f}_{B,2}$ are just finite linear combinations of the coefficients of $\tilde{f}_{B,2}$. We define

$$\tilde{m}_{2,k}(z) = z^{1/2} \cdot m_{1,k}(z)$$

and get that

$$f_2(z) = \sum_{k=0}^\infty \hat{f}_{B,2}^{(k)} \tilde{m}_{2,k}(z) \ , \tag{3.14}$$

at least formally. We next give the necessary bounds to make this rigorous. First we have by definition

$$\hat{f}_{B,2}^{(k)} = \frac{1}{2\pi i} \oint_{|t|=\rho<1} \tilde{f}_{B,2}(-\beta \cdot \log(1-t)) \frac{dt}{t^{k+1}}$$

and therefore

$$|\hat{f}_{B,2}^{(k)}| \leq \frac{1}{\rho^k} \cdot \sup_\varphi |\tilde{f}_{B,2}(-\beta \cdot \log(1 - \rho e^{i\varphi}))| \tag{3.15}$$

and we have to establish a bound for $\tilde{f}_{B,2}$. First note that the function $f_{B,2}(-\beta \cdot \log(1 - t))$ is analytic in the unit disk, and that

$$\frac{d}{dt} f_{B,2}(-\beta \cdot \log(1 - t)) = f'_{B,2}(-\beta \cdot \log(1 - t)) \cdot \beta \cdot \frac{1}{1-t} . \tag{3.16}$$

Furthermore, we note that

$$|\tilde{f}_{B,2}(t)| \leq \frac{1}{\pi^{1/2}} \left(\int_0^\infty \frac{dx}{(1+x)^2 \sqrt{x}} \right) \sup_{x \geq 0} |f'_{B,2}(\frac{t}{1+x})|$$

$$= \frac{\pi^{1/2}}{2} \cdot \sup_{x \geq 0} |f'_{B,2}(\frac{t}{1+x})| . \tag{3.17}$$

We choose now $\hat{\rho}$; $\rho < \hat{\rho} < 1$, and combining the bounds (3.15) and (3.17) with (3.16), we get

$$|\hat{f}_{B,2}^{(k)}| \leq \frac{\pi^{1/2}}{2} \cdot \frac{1}{\rho^k} \cdot \sup_\varphi \cdot \sup_x |f'_{B,2}(-\frac{\beta}{1+x} \cdot \log(1 - \rho e^{i\varphi}))|$$

$$= \frac{\pi^{1/2}}{2} \cdot \frac{1}{\rho^k} \cdot \frac{1}{\beta} \cdot \sup_\varphi |[f_{B,2}(-\beta \cdot \log(1 - z))]'|_{z=\rho e^{i\varphi}} (1-\rho e^{i\varphi})|$$

$$\leq \frac{\pi^{1/2}}{2} \cdot \frac{1}{\beta} \cdot \frac{1+\rho}{\rho^k} \cdot \sup_\varphi |[f_{B,2}(-\beta \cdot \log(1 - z))]'|_{z=\rho e^{i\varphi}} |$$

$$\leq \frac{\pi^{1/2}}{2} \cdot \frac{1}{\beta} \cdot \frac{1+\rho}{\rho^k} \frac{1}{\hat{\rho} - \rho} \sup_\varphi |f_{B,2}(-\beta \cdot \log(1 - \hat{\rho} e^{i\varphi}))| . \tag{3.18}$$

We use now the bound (3.4) and use that

$$|t|^{1/2} \leq a \cdot \exp(\frac{|t|}{2a^2 e}) , \quad \text{for all} \quad a > 0$$

to show that

$$|\hat{f}_{B,2}^{(k)}| \leq \pi^{1/2} \cdot \frac{1}{\beta} \cdot \frac{1}{\rho^k} \frac{1}{\hat{\rho}-\rho} C \cdot a \cdot \exp(-\frac{\beta}{R''} \cdot \log(1 - \hat{\rho})) \qquad (3.19)$$

where $\frac{1}{R''} = \frac{1}{R} + \frac{1}{2a^2e}$. We optimize this by choosing

$$\hat{\rho} = \frac{1+\frac{\beta}{R''}\rho}{1+\frac{\beta}{R''}} ,$$

$$\rho = 1 - \frac{\beta/R''}{k+\beta/R''}$$

and a bound similar to (3.10) can be established, which proves
(3.14) for z in the disk $D_{R''} = \{z \in \mathbb{C} \mid \text{Re}(\frac{1}{z}) > \frac{1}{R'''}\}$ with
$R''' = (R''^{-1} + \beta^{-1})^{-1}$. This completes our discussion of Loeffel's
method.

4. Computer-assisted proofs

4.1. Introduction

We are using the notion of "computer-assisted proof" in the following way : We intend to prove a theorem by invoking a very common mathematical tool, namely the contraction mapping principle in a Banach space. It is well known that this amounts to prove two things for a certain operator K :

- There is an approximate fixed point f_o for which

$$\| f_o - K f_o \| < \varepsilon \ . \tag{4.1.1}$$

- For every f for which $\| f - f_o \| < \beta$ one has the bound on the tangent map DK_f of K at f

$$\| DK_f \| \leq \rho < 1 \ . \tag{4.1.2}$$

If the numbers ε, β, ρ satisfy the inequality

$$\beta \cdot (1 - \rho) > \varepsilon$$

then there is a function $f, \| f - f_o \| < \dfrac{\varepsilon}{1-\rho} \ ,$ for which

$$K f = f$$

and this function is unique among those for which $\| f - f_o \| < \beta$.

We use the computer to prove mathematically rigorous bounds
of the type (4.1.1) and (4.1.2).

To avoid any possible confusion, let us stress that we do not
use the computer to "invent" a proof ("artificial intelligence"),
nor is the program presented here to be considered as a numerical
experiment. Of course, we have done extensive numerical experiments
to guess the "correct" choices of domains of analyticity, the ini-
tial guess of f_0, of ε, β, ρ and the like, but in the program
we omit most of these calculations, which - in our eyes - are
not as interesting as the rigorous computer-produced bounds themselves.

We have stressed in great detail the philosophical implica-
tions and problems of the method in our paper (with H. Koch) [16]
Here, we concentrate on the technical aspects.

The structure of the proof consists in giving bounds, and
the nature of these bounds can be classified as follows.

1)   Consider the space of analytic functions f on the unit
     disk for which the expression

$$|f|_1 = \sum_{n=0}^{\infty} |f_n| \quad \text{is finite}$$

(here $f(z) = \sum_{n=0}^{\infty} f_n z^n$).

On this space we shall consider _balls_ of functions. We shall give balls in which the sum, product,... of functions in two given balls lie.

2) In order to do the above estimates, we need, among other things, interval arithmetics. (I.e. we shall give intervals in which the sum, product,... of points in two given intervals lie.)

3) In order to do the above estimates on a computer, we need some information on the arithmetic as done by a computer. E.g. the sum of two numbers each of which can be represented on the computer may not be representable on the computer. We shall then need an upward- or downward-rounded representable result.

We choose to start the discussion on the lowest possible level, the computer, and we will work our way upward through the points 3), 2), 1) in this order. Parts of this discussion parallel the one given in our joint work with H. Koch [16], and of course, we are strongly dependent on the original ideas of Lanford [32].

4.2. Representable numbers and computer arithmetic

Doing arithmetics on a computer is complicated by the fact that a computer can only represent a finite number of real numbers (in general a subset of the dyadic rationals). We shall now describe some concepts which take this defect into account, and give a reasonably general description of how modern computers do arithmetics (our calculations have been done on an UNIVAC 1100/61).

We assume that integer operations on indices, counting loops, and the like are performed without any rounding errors on the computer, and we do not discuss this matter any further. Let us consider next real numbers. As we have just said, the computer can only represent a finite subset of $\mathbb{R}$, which we call the representable numbers, or reps. We denote the set of reps by R, and we enumerate now a set of requirements a computer must fulfill in order for our method to work. These requirements (or very simple variants thereof) seem to be satisfied on most modern computers.

R1) R contains zero and one.

R2) If $x \in R$, then $-x \in R$.

R3) There is a smallest positive number, x_o, in R.

The arithmetic operations should then satisfy :

R4) If $x \in R$, the computer correctly finds $-x \in R$.

R5) For any pair $x, y \in R$, the computer can correctly decide whether $x = y$, $x > y$, or $x < y$.

R6) Given $x \in R$, and defining $R_e = R \cup$ "undefined", it is possible to define operations up, down : $R \to R_e$ by

$$
\text{up}(x) = \begin{cases} \inf\{y \in R \mid y > x\} & \text{if this set is not empty} \\ \\ \text{undefined otherwise ,} \end{cases}
$$

$$\text{down}(x) = \begin{cases} \sup\{y \in R \,|\, y < x\} & \text{if this set is not empty} \\[2ex] \text{undefined otherwise.} \end{cases}$$

R7) Given $x, y \in R$, the computer can perform their approximate product, denoted $x *_c y$, yielding a result $z \in R_e$ (undefined only if the product is too large or too negative). The precision of the result is such that

$$\text{down}(x *_c y) \leq xy \leq \text{up}(x *_c y) \,.$$

Mutatis mutandis, the same requirements are made for the operation of taking inverses, denoted $1 \div_c x$.

R8) We require for the computer addition

$$\text{down}(x +_c y) \leq x + y \leq \text{up}(x +_c y) \,.$$

In short, <u>the computer is required to perform the arithmetic operations to as many places as is possible, with a minimal rounding error, among the numbers it can represent.</u>

The reader not familiar with programming or not interested in the detailed implementation of R5), R6) can skip the remainder of this subsection.

<u>Notation</u> : We denote $x +_u y = \text{up}(x +_c y)$, $x *_u y = \text{up}(x *_c y)$, and similarly $x +_d y = \text{down}(x +_c y)$ etc.

All reps which we use in our program have a name starting
with r... . Integers have names starting with I, J, K, L, M
or N.

The following subroutine initializes some reps (and for later
use, some scalars (see below), and the degree K of polynomial
approximation).

```
C   Initialization
C   ==============

      SUBROUTINE INITIALIZE(K)
      INCLUDE BIGN.STUFF
      COMMON/SWITCH/ITEST
      ITEST=1
      CALL DEG(K)
      NERR =NDEG
      rZERO =0.D+00
      rONE  =1.D+00
      rTWO  =2.D+00
      rTHREE=3.D+00
      rFOUR =4.D+00
      rHALF =.5D+00
      r3HALF=1.5D+00
      sZERO =sCONST(rZERO)
      sONE  =sCONST(rONE)
      sTWO  =sCONST(rTWO)
      sTHREE=sCONST(rTHREE)
      sFOUR =sCONST(rFOUR)
      sHALF =sCONST(rHALF)
      s3HALF=sCONST(r3HALF)
      RETURN
      END
```


See Section 5.5   for the file BIGN.STUFF.

We now describe in detail the set R for a Univac, with the
FORTRAN-ASCII compiler. R coincides with what is conventionally
called "double reals". Double real numbers consist of two words
of 36 bits each, which we number from 1 to 72 from left to right.
If all 72 bits are 0, the number is interpreted as $0 \in \mathbb{R}$.   All

positive reps have a 0 bit in position 1 and a 1 bit in position

13. The number 1/2 is in R and is represented as

200040000000, 000000000000

in octal notation. In fact, the rules are as follows. Bits 2 to

12 represent the exponent and bits 13 to 72 the mantissa. View

bits 2 to 12 as the binary representation of a non-negative in-

teger, n, and bits 13 to 72 as the fractional part $x \in [1/2,1)$

of a binary number. Then the number represented by the 72 bits is

$$x2^{n-2^{10}} .$$

Thus, 1/2 is represented as $0.5 \times 2^{2^{10}-2^{10}}$. The smallest pos-

itive rep, i.e. x_0, is

$$000040000000, \; 000000000000 = 0.5 \cdot 2^{-2^{10}} \sim 0.278 \cdot 10^{-308}$$

and the largest positive rep is

$$677777777777, \; 777777777777 = (1-2^{-60}) \cdot 2^{(2^{11}-1)-2^{10}} \sim 0.899 \cdot 10^{+308} .$$

All negative numbers x are represented as the two's complement

of -x.

It is easy to see that this set satisfies R1)-R3), and R4).

The condition R5) is <u>not</u> directly implemented on the Univac, (a

mistake in our paper [16]) because in fact the internal maximum function DMAX1 does not work properly when underflow occurs. Thus it has to be replaced by some (obvious) handwritten code. BITS (K,L,M) reads or sets bits L,...,L + M - 1 in the word K.

```
      REAL FUNCTION rMAX2*8(r1,r2)
      INCLUDE BIGN.STUFF
      EQUIVALENCE(rI1,K1)
      EQUIVALENCE(rI2,K2)
C..try normal comparison first
      IF(r1-r2)1,3,2
    1 CONTINUE
      rMAX2=r2
      RETURN
    2 CONTINUE
      rMAX2=r1
      RETURN
C..seem to be equal
    3 CONTINUE
      IF(r1)4,1,5
C..both numbers are negative
    4 CONTINUE
      rI1=-r1
      rI2=-r2
C..if both numbers are different from zero and their difference
C  is zero then the exponents must have a 0 bit in the leading
C  position. We multiply them by 2**10 by setting this bit one.
C  For such numbers underflow cannot occur!
      BITS(K1,2,1)=1
      BITS(K2,2,1)=1
      IF(rI2-rI1)1,1,2
C..similar procedure if both positive
    5 CONTINUE
      rI1=r1
      rI2=r2
      BITS(K1,2,1)=1
      BITS(K2,2,1)=1
      IF(rI1-rI2)1,1,2
      END
```

We also need the maximum of 3 arguments and the minimum of 2.

```
REAL FUNCTION rMAX3*8(r1,r2,r3)
INCLUDE BIGN.STUFF
rMAX3=rMAX2(rMAX2(r1,r2),r3)
RETURN
END

REAL FUNCTION rMIN2*8(r1,r2)
INCLUDE BIGN.STUFF
rMIN2=-rMAX2(-r1,-r2)
RETURN
END
```

We next discuss up and down.

Essentially, up consists in adding 1 to bit 72, with the 72 bits viewed as a single huge integer. The only programming precaution, which may be necessary on other computer models as well, is the following. The integer addition of the Univac 1100 does not correctly operate on the numbers 777777777777 and 777777777776 when they are viewed as <u>positive</u> integers, because, internally there is no "unsigned integer type" and the first number is considered to be 0 and the second to be -1, and therefore the first never occurs as the result of an integer operation. We handle these special cases by operating directly on bits.

It is straightforward, if tedious, to implement the operations up, down : $R \to R_e$ in practice. We give below the Fortran code for up on a Univac 1100. Down is realized as

$$\text{down}(x) = -_c \text{up}(-_c x) \ .$$

```
      REAL FUNCTION rUP*8(r)
      INCLUDE BIGN.STUFF
      DIMENSION LL(2)
      EQUIVALENCE (rRES,LL(1),K),(LL(2),L)
      DATA L351/03777777777777/
      rRES=r
C..if low part of mantissa is, 0, 777777777776, or 777777777777
C  then exceptional handling
      IF(BITS(L,1,35).EQ.L351)GOTO 3
      IF(L)2,1,2
C..low part zero. Check if r is zero
    1 CONTINUE
      IF(rRES.NE.0.D+00)GOTO 2
C..r is zero. rUP is smallest positive number
      BITS(K,13,1)=1
      rUP=rRES
      RETURN
C..generic case, add one to lower part of mantissa
    2 CONTINUE
      L=L+1
      rUP=rRES
      RETURN
C..lower part of mantissa is 777777777776 or 777777777777
    3 CONTINUE
      IF(BITS(L,36,1).EQ.1)GOTO 4
C..case of 777777777776, changed to 777777777777
      BITS(L,36,1)=1
      rUP=rRES
      RETURN
C..case of 777777777777
    4 CONTINUE
      L=0
      KSAVE=BITS(K,13,1)
      BITS(K,13,1)=1
      K=K+1
C..check of overflow or underflow
      IF(BITS(K,1,1).EQ.KSAVE)GOTO 5
C..restore bit 13
      BITS(K,13,1)=KSAVE
      rUP=rRES
      RETURN
C..distinguish overflow from underflow
    5 CONTINUE
      IF(KSAVE.EQ.0)GOTO 6
C..overflow
      WRITE(6,*)'overflow in rUP'
      rUP=r
      RETURN
C..underflow, result is zero
    6 CONTINUE
      L=0
      K=0
      rUP=rRES
      RETURN
      END


      REAL FUNCTION rDOWN*8(r)
      INCLUDE BIGN.STUFF
      rDOWN=-rUP(-r)
      RETURN
      END
```

According to the description of the Univac 1100, requirements R7),

R8) seem to be satisfied.

Note that there are more modern designs (e.g. the chip 8087),

in which arithmetic operations can be asked to round upward or

downward so that these questions can be relegated to the hardware.

4.3. Intervals and scalars

We now discuss the arithmetics of intervals, in fact a well-

known subject see e.g. [36]. Operations on intervals are defined in such

a way that the resulting interval contains all possible images of the

source. E.g. in addition, one defines the sum $J_1 + J_2$ of two

intervals J_1, J_2 as the set $\{x_1 + x_2 | x_1 \in J_1, x_2 \in J_2\}$.

On the computer, we have to adapt this idea to the fact that

intervals can only have reps (i.e. elements of R) as their end-

points. We call such closed intervals scalars, and we denote S

the set of all scalars.

Definition : $S = \{[r_1, r_2] | r_1 \in R, r_2 \in R, r_1 \leq r_2\}$.

It is convenient, in Fortran, to accommodate the two endpoints

r_1, r_2 as the real and imaginary part of a complex number. In

Pascal, one would define

    TYPE SCALAR = RECORD L, R : REAL END ;

Before we discuss the operations on scalars, we need to produce them, by a set of rather trivial subroutines which we indicate

for completeness. Note that we adopt the <u>convention</u> that all sca-

lars have a name starting with s... .

4.3.1 <u>Conversions to and from scalars</u>

To a rep r we associate the scalar [r,r]. Some constants

have been given names in the subroutine initialize.

Below, there is a conversion rep → scalar, "arithmetic num-

ber" → scalar, integer → scalar. (Arithmetic numbers are reps ob-

tained through a calculation whose result serves as "numerical

guess" for some rigorous bound (e.g. the domain which will be

chosen for a function, or a starting point for a rigorous Newton

algorithm (see below).)

```
COMPLEX FUNCTION sCONST*16(r)
INCLUDE BIGN.STUFF
sCONST=DCMPLX(r,r)
RETURN
END

COMPLEX FUNCTION saCONST*16(a)
INCLUDE BIGN.STUFF
r=a
saCONST=sCONST(r)
RETURN
END

COMPLEX FUNCTION siCONST*16(I)
INCLUDE BIGN.STUFF
siCONST=sCONST(DFLOAT(I))
RETURN
END
```

Of course, we also need to extract the two reps of a scalar. This is

done with the intrinsic functions DREAL, and DIMAG. It is also convenient

to have at our disposal two functions S → R : If s∈S then

$$rMAXABS(s) = \max_{x \in s} |x| \; ,$$

$$rMINABS(s) = \min_{x \in s} |x| \; .$$

```
      REAL FUNCTION rMAXABS*8(s)
      INCLUDE BIGN.STUFF
C..max /x/ , when x in s
      rMAXABS=rMAX2(-DREAL(s),DIMAG(s))
      RETURN
      END


      REAL FUNCTION rMINABS*8(s)
      INCLUDE BIGN.STUFF
C..min /x/ , when x in s
      rMINABS=rMAX3(rZERO,DREAL(s),-DIMAG(s))
      RETURN
      END
```

Notation : We write $\max_r |s|$ for rMAXABS(s) and $\min_r |s|$ for rMINABS(s).

4.3.2 Functions of scalars

We next list the simplest functions on scalars : $S \to S$. sNEG implements $x \to -x$; i.e. if $s \in S$ then for every $x \in s$, one has $-x \in sNEG(s)$. Evidently if $s = [r_1, r_2]$ then $sNEG(s) = [-_c r_2 , -_c r_1]$.

```
      COMPLEX FUNCTION sNEG*16(s)
      INCLUDE BIGN.STUFF
      sNEG=DCMPLX(-DIMAG(s),-DREAL(s))
      RETURN
      END
```

The next three functions extract the absolute value, and the upper and lower bounds : If $s \in S$, then

$$sABS(s) = [\min_{x \in s} |x|, \max_{x \in s} |x|] \; ,$$

$$
\text{sUPPER}(s) = \begin{cases} \text{error if} \quad (\inf_{x \in s} x) < 0 \qquad \text{*)} \\[3mm] [\max_{x \in s} x, \max_{x \in s} x] \quad \text{otherwise} \quad , \end{cases}
$$

$$
\text{sLOWER}(s) = \begin{cases} \text{error if} \quad (\inf_{x \in s} x) < 0 \\[3mm] [\min_{x \in s} x, \min_{x \in s} x] \quad \text{otherwise} \quad . \end{cases}
$$

(The last two functions will be useful in establishing upper and lower bounds while checking for non-negativity.)

```
      COMPLEX FUNCTION sABS*16(s)
      INCLUDE BIGN.STUFF
      rL=DREAL(s)
      rU=DIMAG(s)
      sABS=DCMPLX(rMAX3(rZERO,rL,-rU),rMAX2(-rL,rU))
      RETURN
      END
```

```
      COMPLEX FUNCTION sUPPER*16(s)
      INCLUDE BIGN.STUFF
C..a scalar equal to upper end of s
C   s should be non-negative
      IF(DREAL(s).LT.rZERO)THEN
         WRITE(6,*)´error in sUPPER´,s
         SUPPER=sCONST(rMAXABS(s))
         RETURN
      ENDIF
      sUPPER=sCONST(DIMAG(s))
      RETURN
      END
```

---

*) most of our programs attempt some sort of recovery in case of errors, but the subsequent bounds are wrong.

```
      COMPLEX FUNCTION sLOWER*16(s)
      INCLUDE BIGN.STUFF
C..s should be non-negative
      IF(DREAL(s).LT.rZERO)THEN
        WRITE(6,*)´error in sLOWER´,s
        sLOWER=sCONST(rMAXABS(s))
        RETURN
      ENDIF
      sLOWER=sCONST(DREAL(s))
      RETURN
      END
```

<u>Notation</u> : We write $\max_s(s)$ for sUPPER(s), $\min_s(s)$ for

sLOWER(s) and $|s|$ for sABS(s).

We next consider the simple arithmetic functions : Inter-

section, addition, negative, subtraction, multiplication, inverse,

quotient. The programs should be almost self-explanatory. Let

$s_1, s_2 \epsilon S$, $s_1 = [r_{1L}, r_{1U}]$, $s_2 = [r_{2L}, r_{2U}]$. Then

$$
\text{sINTER}(s_1, s_2) = \begin{cases} \text{error if } \{x|x\epsilon s_1\} \cap \{x|x\epsilon s_2\} = \emptyset \\ \\ \{x|x\epsilon s_1 \text{ and } x\epsilon s_2\} \end{cases}
$$

sSUM(s_1, s_2) defines a scalar which contains

$\{x = x_1 + x_2 | x_1 \epsilon s_1, x_2 \epsilon s_2\}$. The easiest definition of sSUM would

be sSUM$(s_1, s_2) = [\text{rDOWN}(r_{1L} +_c r_{2L}), \text{rUP}(r_{1U} +_c r_{2U})]$. We choose

a more complicated procedure to make [0,0] a neutral element

for addition. Similarly,

$$\{x = x_1 - x_2 \mid x_1 \epsilon s_1 , x_2 \epsilon s_2\} \subset \text{sDIFF}(s_1, s_2) .$$

In the product we have distinguished the nine relative possible

signs of the endpoints and we enforce [0,0] to yield a product

[0,0] (but, to save time, [1,1] is not a neutral element)

$$sINV(s_1) = \begin{cases} \text{error if } 0 \in s_1 \\ \\ [rDOWN(1 \div_c r_{1U}), rUP(1 \div_c r_{1L})] \end{cases} .$$

```
      COMPLEX FUNCTION SINTER*16(s1,s2)
C..computes intersection of two scalars
      INCLUDE BIGN.STUFF
      EQUIVALENCE(rIL,KL)
      EQUIVALENCE(rIU,KU)
      rL=rMAX2(DREAL(s1),DREAL(s2))
      rU=rMIN2(DIMAG(s1),DIMAG(s2))
      IF(rL-rU)1,3,2
    2 CONTINUE
      WRITE(6,*)'empty intersection',s1,s2
      SINTER=s1
      RETURN
    1 CONTINUE
      SINTER=DCMPLX(rL,rU)
      RETURN
    3 CONTINUE
      IF(rL)4,1,5
    4 CONTINUE
      rIL=-rL
      rUL=-rU
      BITS(KL,2,1)=1
      BITS(KU,2,1)=1
      IF(rIU-rIL)1,1,2
    5 CONTINUE
      rIL=rL
      rUL=rU
      BITS(KL,2,1)=1
      BITS(KU,2,1)=1
      IF(rIL-rIU)1,1,2
      END


      COMPLEX FUNCTION sSUM*16(s1,s2)
      INCLUDE BIGN.STUFF
      r1L=DREAL(s1)
      r1U=DIMAG(s1)
      r2L=DREAL(s2)
      r2U=DIMAG(s2)
      IF    (r1L.EQ.rZERO.AND.r1L.EQ.r1U)THEN
          sSUM=s2
      ELSEIF(r2L.EQ.rZERO.AND.r2L.EQ.r2U)THEN
          sSUM=s1
      ELSE
          sSUM=DCMPLX(rDOWN(r1L+r2L),rUP(r1U+r2U))
      ENDIF
      RETURN
      END
```

```
COMPLEX FUNCTION sDIFF*16(s1,s2)
INCLUDE BIGN.STUFF
sDIFF=sSUM(s1,sNEG(s2))
RETURN
END




COMPLEX FUNCTION sPROD*16(s1,s2)
INCLUDE BIGN.STUFF
r1L=DREAL(s1)
r1U=DIMAG(s1)
r2L=DREAL(s2)
r2U=DIMAG(s2)
IF    ((r1U.EQ.rZERO.AND.r1L.EQ.r1U).OR.
*       (r2U.EQ.rZERO.AND.r2L.EQ.r2U))THEN
  sPROD=sZERO
ELSEIF(r1L.GE.rZERO)THEN
    IF    (r2L.GE.rZERO)THEN
      sPROD=DCMPLX(rDOWN(r1L*r2L),rUP(r1U*r2U))
    ELSEIF(r2U.LT.rZERO)THEN
      sPROD=DCMPLX(rDOWN(r1U*r2L),rUP(r1L*r2U))
    ELSE
      sPROD=DCMPLX(rDOWN(r1U*r2L),rUP(r1U*r2U))
    ENDIF
ELSEIF(r1U.LT.rZERO)THEN
    IF    (r2L.GE.rZERO)THEN
      sPROD=DCMPLX(rDOWN(r1L*r2U),rUP(r1U*r2L))
    ELSEIF(r2U.LT.rZERO)THEN
      sPROD=DCMPLX(rDOWN(r1U*r2U),rUP(r1L*r2L))
    ELSE
      sPROD=DCMPLX(rDOWN(r1L*r2U),rUP(r1L*r2L))
    ENDIF
ELSE
    IF    (r2L.GE.rZERO)THEN
      sPROD=DCMPLX(rDOWN(r1L*r2U),rUP(r1U*r2U))
    ELSEIF(r2U.LT.rZERO)THEN
      sPROD=DCMPLX(rDOWN(r1U*r2L),rUP(r1L*r2L))
    ELSE
      sPROD=DCMPLX(rDOWN(rMIN2(r1L*r2U,r1U*r2L)),
*                  rUP  (rMAX2(r1U*r2U,r1L*r2L)))
    ENDIF
ENDIF
RETURN
END




COMPLEX FUNCTION sINV*16(s)
INCLUDE BIGN.STUFF
LOGICAL logZERO
IF(logZERO(s))THEN
  WRITE(6,*)'error in sINV',s
  sINV=s
  RETURN
ENDIF
rL=DREAL(s)
rU=DIMAG(s)
sINV=DCMPLX(rDOWN(rONE/rU),rUP(rONE/rL))
RETURN
END
```

```
COMPLEX FUNCTION sQUOT*16(s1,s2)
INCLUDE BIGN.STUFF
sQUOT=sPROD(s1,sINV(s2))
RETURN
END
```

For s∈S and K an integer, we define sPOWER(s) ⊃ {x^K, x∈S, if

K ≥ 0, error otherwise}. (In other words, successive multiplica-

tion.) The procedure is based on a binary decomposition of K.

```
COMPLEX FUNCTION sPOWER*16(s,K)
INCLUDE BIGN.STUFF
IF(K.LT.0)THEN
   WRITE(6,*)´error in sPOWER, K not positive´
   STOP
ENDIF
IF(rMAXABS(s).EQ.rZERO)GOTO 1
M=K
s1=s
sT1=sONE
3 CONTINUE
IF(M.EQ.0)GOTO 4
MM=M/2
IF(MM+MM.EQ.M)GOTO 2
sT1=sPROD(sT1,s1)
M=M-1
GOTO 3
2 CONTINUE
s1=sPROD(s1,s1)
M=MM
GOTO 3
1 CONTINUE
sT1=sZERO
IF(K.EQ.0)sT1=sONE
4 CONTINUE
sPOWER=sT1
RETURN
END
```

We also need trigonometric functions, square roots, exponentials

and logarithms. The routines provided by Univac clearly do not

have asserted error bounds (in contrast to simple arithmetics)

and so we write our own.

We write e^x as a polynomial of degree N (= 70) and bound the remainder, when $1 \geq |x|$ by

$$| \sum_{n=N+1}^{\infty} x^n/x! | \leq \frac{|x|^{N+1}}{(N+1)!}/(1 - \frac{|x|}{N+2}).$$

If $\max_r x < 0$ we compute $1/e^{-x}$ and if $|x| > 1$ we divide x sufficiently often by 2, $y = x/2^P$, $1 \geq |y|$ and compute $e^x = (e^y)^{2^P}$.

```
COMPLEX FUNCTION sEXP*16(s)
INCLUDE BIGN.STUFF
PARAMETER IMAX=70
DIMENSION sFACT(IMAX+1)
sT1=sONE
DO 1 I=1,IMAX+1
sFACT(I)=sT1
sT1=sPROD(sT1,siCONST(I+1))
1 CONTINUE
IF(DIMAG(s).LT.rZERO)THEN
   INEG=1
   sT1=sNEG(s)
ELSE
   INEG=0
   sT1=s
ENDIF
IREDUCE=0
2 CONTINUE
IF(rMAXABS(sT1).GT.rONE)THEN
   sT1=sQUOT(sT1,sTWO)
   IREDUCE=IREDUCE+1
   GOTO 2
ENDIF
sT2=sQUOT(sPOWER(sT1,IMAX+1),sFACT(IMAX+1))
sT3=sDIFF(sONE,sQUOT(sABS(sT1),siCONST(IMAX+2)))
rT2=rMAXABS(sQUOT(sABS(sT2),sT3))
sT2=DCMPLX(-rT2,rT2)
DO 3 I=1,IMAX
K=IMAX+1-I
sT2=sSUM(sT2,sQUOT(sPOWER(sT1,K),sFACT(K)))
3 CONTINUE
sT2=sSUM(sT2,sONE)
DO 4 I=1,IREDUCE
sT2=sPROD(sT2,sT2)
4 CONTINUE
IF(INEG.EQ.1)sT2=sINV(sT2)
sEXP=sT2
RETURN
END
```

The logarithm is the first interesting problem. We view the logarithm y of x as the solution of $e^y - x = 0$. We do an interval version of the Newton algorithm which we describe in detail now.

If, generally, we have to solve $f(z) = 0$ where f is from S to S, then we first determine by a non-rigorous calculation an approximate zero $z_0 \in R$. We then set $s_1 = [z_0, z_0]$, and determine $\varepsilon = f(s_1)$ (note $\varepsilon \in S$!). If $0 \notin \varepsilon$ we enlarge ε to contain 0. We now choose $\beta \in S$, e.g. $\beta = [-10^{-3}, 10^{-3}]$ and we compute $f'(\beta + s_1) = d$ (again in S). Note that for every $x \in s_1 + \beta$ we find $f'(x) \in d$. Therefore, we find that <u>if</u> $-\varepsilon/d \in S$ <u>is contained in</u> β <u>then</u> f <u>must have a root in</u> $s_1 - \varepsilon/d$. We can iterate this procedure and in general the interval in which the root is to be found will shrink.

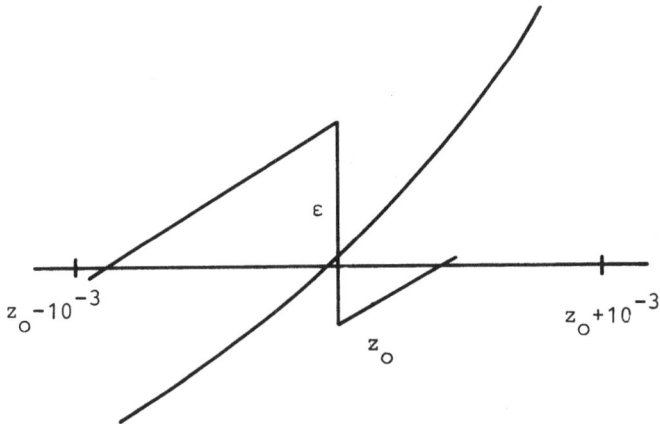

<u>Fig. 4.1</u>. The interval $[z_0 - 10^{-3}, z_0 + 10^{-3}]$ is contracted into itself.

<u>Remark</u> : The above remarks apply as well if f <u>is in a set</u> of functions. E.g. if we compute $e^y - s = 0$, $s \in S$ then $f(y) = e^y - s$, for $y \in S$. We choose below z_0 as the approximate zero for the

center of the interval s. Of course, if s is very wide
then -ε/d will not be contained in β and the algorithm does
not allow to find a root, and in all cases the "root" will be an
interval whose size must increase with the diameter of s.

```
      COMPLEX FUNCTION sLOG*16(s)
      INCLUDE BIGN.STUFF
C..sT1 is approximate value
      sT1=saCONST(DLOG(aCENT(s)))
C..compute value at central guess
      sEPSILON=sDIFF(sEXP(sT1),s)
C..we enlarge epsilon to contain 0
      sEPSILON=DCMPLX(rMIN2(DREAL(sEPSILON),rZERO)
     *                ,rMAX2(DIMAG(sEPSILON),rZERO))
C..Newton for intervals
      rBETA=1.D-3
      sBETA=DCMPLX(-rBETA,rBETA)
      DO 1 I=1,5
      sINT=sSUM(sT1,sBETA)
      sDER=sEXP(sINT)
      sNEW=sNEG(sQUOT(sEPSILON,sDER))
      IF(sINTER(sNEW,sBETA).NE.sNEW)THEN
        WRITE(6,*)´error in sLOG, no contraction´
        sLOG=sINT
        RETURN
      ENDIF
      sBETA=sNEW
    1 CONTINUE
      sLOG=sSUM(sT1,sBETA)
      RETURN
      END
```

(See below for the definition of aCENT.) It is now evident how
to compute $(s_1)^{s_2}$, when $s_1,s_2 \in S$:

```
      COMPLEX FUNCTION sTOs*16(s1,s2)
C..sTOs=s1**s2
      INCLUDE BIGN.STUFF
      sTOs=sEXP(sPROD(sLOG(s1),s2))
      RETURN
      END
```

The square root is computed by the Newton algorithm.

```
        COMPLEX FUNCTION sSQRT*16(s)
        INCLUDE BIGN.STUFF
C..choose scaling factor for below
        sFACT =saCONST(1.D+5)
        sFACT2=sPOWER(sFACT,2)
C..treat first special cases
        IF(DREAL(s).LT.rZERO)THEN
          WRITE(6,*)'error in sSQRT, argument negative',s
          sSQRT=sZERO
          RETURN
        ELSEIF(DIMAG(s).EQ.rZERO)THEN
          sSQRT=sZERO
          RETURN
        ENDIF
        IZERO=0
        sT1=s
        IF(DREAL(sT1).EQ.rZERO)THEN
          IZERO=1
          sT1=sUPPER(sT1)
        ENDIF
        ISMALL=0
        ITOSMALL=0
        IF(rMAXABS(sT1).LT.rONE)THEN
          ISMALL=1
   1      CONTINUE
          IF(rMAXABS(sT1).LT.1.D-11)THEN
            ITOSMALL=ITOSMALL+1
            sT1=sPROD(sT1,sFACT2)
            GOTO 1
          ENDIF
          sT1=sINV(sT1)
        ENDIF
C..sT2 is approximate value
        sT2=saCONST(DSQRT(aCENT(sT1)))
C..compute value at central guess
        sEPSILON=sDIFF(sPROD(sT2,sT2),sT1)
C..we increase epsilon to contain 0
        sEPSILON=DCMPLX(rMIN2(DREAL(sEPSILON),rZERO)
     *                  ,rMAX2(DIMAG(sEPSILON),rZERO))
C..Newton for intervals
        rBETA=1.D-3
        sBETA=DCMPLX(-rBETA,rBETA)
        DO 2 I=1,5
        sINT=sSUM(sT2,sBETA)
        sDER=sSUM(sINT,sINT)
        sNEW=sNEG(sQUOT(sEPSILON,sDER))
        IF(sINTER(sNEW,sBETA).NE.sNEW)THEN
          WRITE(6,*)'error in sSQRT, no contraction.',I,s
          sT2=sINT
          IF(ISMALL.EQ.1)sT2=sINV(sT2)
          IF(IZERO.EQ.1)sT2=DCMPLX(rZERO,DIMAG(sT2))
          sSQRT=sT2
          RETURN
        ENDIF
        sBETA=sNEW
   2    CONTINUE
```

```
    sT2=sSUM(sT2,sBETA)
    IF(ISMALL.EQ.1)THEN
      sT2=sINV(sT2)
      DO 3 I=1,ITOSMALL
      sT2=sQUOT(sT2,sFACT)
  3   CONTINUE
    ENDIF
    IF(IZERO.EQ.1)sT2=DCMPLX(rZERO,DIMAG(sT2))
    sSQRT=sT2
    RETURN
    END
```

Finally, to compute the sine, we sum the series to a degree
2m and 2m + 1. Since the series is alternating, this provides
upper and lower bounds, if $|x| < \pi$ and m is large (49 is
enough). Cosine and tangent are considered as derived quantities.

```
    COMPLEX FUNCTION sSIN*16(s)
    INCLUDE BIGN.STUFF
    PARAMETER IMAX=49
    DIMENSION sFAC(IMAX)
    IF(rMAXABS(s).GE.rMAXABS(siCONST(IMAX)))THEN
      WRITE(6,*)'error in sSIN, argument to large',s
    ENDIF
    sT1=sONE
    DO 1 I=1,IMAX
    sFAC(I)=sT1
    sT1=sPROD(sT1,siCONST(I+1))
  1 CONTINUE
    sT1=sQUOT(sPOWER(s,IMAX),sFAC(IMAX))
    sT2=sZERO
    sT3=sONE
    DO 2 I=2,IMAX-1,2
    sT3=sNEG(sT3)
    sT1=sSUM(sT1,sPROD(sT3,sQUOT(sPOWER(s,IMAX-I),sFAC(IMAX-I))))
    sT2=sSUM(sT2,sPROD(sT3,sQUOT(sPOWER(s,IMAX-I),sFAC(IMAX-I))))
  2 CONTINUE
    sSIN=DCMPLX(rMAX2(rMIN2(DREAL(sT1),DREAL(sT2)),-rONE)
   *           ,rMIN2(rMAX2(DIMAG(sT1),DIMAG(sT2)), rONE))
    RETURN
    END


    COMPLEX FUNCTION sCOS*16(s)
    INCLUDE BIGN.STUFF
    IF(rMAXABS(s).GE.rMINABS(sQUOT(sPI,sTWO)))THEN
      WRITE(6,*)'error in sCOS, argument to large',s
    ENDIF
    sCOS=sSQRT(sDIFF(sONE,sPOWER(sSIN(s),2)))
    RETURN
    END
```

```
COMPLEX FUNCTION sTAN*16(s)
INCLUDE BIGN.STUFF
sTAN=sQUOT(sSIN(s),sCOS(s))
RETURN
END
```

Remark : The above subroutines may give an error indication although it would strictly speaking be possible to find an answer,
(e.g. the root of a very large number). However, since we did not
need the answers in such cases, they have not been implemented
and yield an error message instead.

In order to present the full program, we list here two further functions which are used.

```
      LOGICAL FUNCTION logZERO(s)
C..returns true if s contains 0,false otherwise
      INCLUDE BIGN.STUFF
      logZERO=DREAL(s).LE.rZERO .AND. DIMAG(s).GE.rZERO
      RETURN
      END
```

```
      REAL FUNCTION aCENT*8(s)
      INCLUDE BIGN.STUFF
      aCENT=(DIMAG(s)+DREAL(s))*rHALF
      RETURN
      END
```

Notation : In subsequent discussions, for $s_1, s_2 \in S$, $s_1 \cdot_s s_2$,
$s_1 +_s s_2$ etc, denote the action of the corresponding subroutines;
that is, we indicate explicitly in which sense the "+" sign is
to be understood.

4.4. Neighborhood arithmetics in function space

The basic idea is to define standard balls in a function space (essentially polynomials with coefficients in S, and a higher order term) and to show that the necessary operations lead to results in S or in a standard ball. Our estimates are again a compromise between efficiency and the quality of the bound.

4.4.1 The standard balls, "vectors"

We shall consider analytic functions on the unit disk $D = \{z \in \mathbb{C} \mid |z| < 1\}$. If we denote $f(z) = \sum_{n=0}^{\infty} f_n z^n$ the Taylor expansion of such functions at zero, then we shall restrict our attention to the set A of functions for which

i) $f_n \in \mathbb{R}$, $n = 0,1,2,\ldots$,

ii) $\sum_{n=0}^{\infty} |f_n| < \infty$.

If we denote by $|f|_1$ the above sum, then A, equipped with the norm $|.|_1$ is a (real) Banach space.

In order to develop a neighborhood arithmetic we have to define the analog of intervals : Standard Balls. We choose, once and for all, a degree $N > 0$ and we then define a standard ball of degree N by giving $N + 1$ scalars (intervals) $v_i \in S$, $i = 0,\ldots,N$ and two non-negative reps (numbers) $v^{(G)} \in \mathbb{R}$ and $v^{(H)} \in \mathbb{R}$. We denote this vector $\underline{v} = \{v_0,\ldots,v_N, v^{(G)}, v^{(H)}\}$, and the associated ball in A is defined by

$$A(\underline{v}) = \{f \in A \mid f = \sum_{i=0}^{N} f_i z^i + f_G(z) + f_H(z) ,$$

$$f_i \in v_i \quad , \quad i = 0, 1, \ldots, N \quad ,$$

$$|f_G|_1 \leq v^{(G)} \quad , \quad |f_H|_1 \leq v^{(H)} \quad ,$$

$$f_G(z) = \mathcal{O}(z) \quad \text{as} \quad z \to 0 \quad ,$$

$$f_H(z) = \mathcal{O}(z^{N+1}) \quad \text{as} \quad z \to 0 \} \quad .$$

We call f_G the "general" and f_H the "higher order" part of f.
The choice $f_G(z) = \mathcal{O}(z)$ may seem artificial, and $f_G(z) = \mathcal{O}(1)$
is perhaps more natural, but in some of our estimates our choice
will turn out to be more convenient.

$A(\underline{v})$ has a natural decomposition in a polynomial, general
and higher order part. We call B the set of $\underline{v}$ as described
above. We call $\underline{v} \in B$ a vector and in the program, we denote elements in B by $v\ldots$. We again list, in order of increasing
complexity, the various operations on vectors.

4.4.2 Definition of data structures

A vector is represented as a FORTRAN-array. The correspondence
of $\underline{v} \in B$ with an array, say vA, is

$$vA(i+1) = v_i \quad , \quad i = 0, \ldots, N \quad ,$$

i.e. $vA(i+1)$ is a complex number whose two components represent
the scalar $v_i \in S$. There is an integer $NERR > N + 1$ and

$$vA(NERR) = [-_c\ v^{(G)}, v^{(H)}]\ .$$

The data structure would be more logical in a language like

Pascal, where

```
type interval=record
        lower:real;
        upper:real
     end;
type polynomial=array[0..ndeg]of interval;
```

```
type ball=record
        x:polynomial;
        g:real;
        h:real
     end;
```

Also, in a language like ADA, operations on vectors could be
called with infix notation, while we have to use explicit subroutine calls. There is a set of subroutines to extract and set
the error term, and also a subroutine to fix the degree N. The
polynomial part of a vector is denoted x,... and there are
also subroutines to form vectors from polynomials and errors
(which are denoted e...).

```
C   Set degree
C   ==========

        SUBROUTINE DEG(K)
        INCLUDE BIGN.STUFF
        N =K
        N1=N+1
        N2=N1+1
        RETURN
        END


C   Operations acting on errors
C   ===========================

C   Extraction
C   ----------
        REAL FUNCTION rG*8(v)
        INCLUDE BIGN.STUFF
C..returns bound on general of v
        DIMENSION v(1)
        rG=-DREAL(v(NERR))
        RETURN
        END
```

```
      REAL FUNCTION rH*8(v)
      INCLUDE BIGN.STUFF
C..returns bound on higher order of v
      DIMENSION v(1)
      rH=DIMAG(v(NERR))
      RETURN
      END


C  Recombination
C  -------------
      COMPLEX FUNCTION error*16(rG,rH)
      INCLUDE BIGN.STUFF
      error=DCMPLX(-rG,rH)
      RETURN
      END


C  Addition of two error bounds
C  ----------------------------
C  (Zero is a neutral element)
      COMPLEX FUNCTION eSUM*16(e1,e2)
      INCLUDE BIGN.STUFF
      s1G=sCONST(-DREAL(e1))
      s1H=sCONST( DIMAG(e1))
      s2G=sCONST(-DREAL(e2))
      s2H=sCONST( DIMAG(e2))
      eSUM=error(DIMAG(sSUM(s1G,s2G)),DIMAG(sSUM(s1H,s2H)))
      RETURN
      END




C     Make vectors
C     ------------
      SUBROUTINE FxeTOv(x,e,v)
C..combine power series and error into vector
      INCLUDE BIGN.STUFF
      DIMENSION x(1),v(1)
      DO 1 I=1,N1
      v(I)=x(I)
    1 CONTINUE
      CALL FCLEAN(v)
      v(NERR)=e
      RETURN
      END
```

```
      SUBROUTINE FCLEAN(v)
C..set unused elements to zero
      INCLUDE BIGN.STUFF
      DIMENSION v(1)
      DO 1 I=N2,NERR-1
      v(I)=sZERO
    1 CONTINUE
      RETURN
      END
```

4.4.3 Forming simple vectors

The following subroutines define certain vectors : FSETZERO(v)
sets v to the zero function FSETLIN(v,sC,sL) forms the linear
function v_0 = sC, v_1 = sL, v_i = [0,0] i = 2,...,N, $v^{(G)}$ = $v^{(H)}$ = 0
(we use in this, and later examples the more readable notation
v(z) = sC + sL·z).

FSETINV(v,s) defines a vector $\underline{v} \in B$ such that $A(\underline{v})$ con-
tains all functions {f(z) = 1/(1 + y·z)|y∈s}. Such a v is
given by

$$v_i = (-s)^i , \quad i = 0,...,N ,$$

$$v^{(G)} = 0 , \quad v^{(H)} \geq |s|^{N+1}/(1 - |s|) .$$

(Note that v can only be defined if |s| < 1.)

FSETLOG(v,s) defines similarly v(z) = log(1 + zs), pro-
vided |s| < 1. The error term is bounded by $v^{(G)}$ = 0,
$v^{(H)} \geq |s|^{N+1}/((N+1)(1-|s|))$.

FSETMSQRT($\underline{v}$,s)   defines $v(z) = (1+zs)^{-1/2}$,   provided

$|s| < 1$.   The error term is bounded by $v^{(G)} = 0$,

$v^{(H)} \geq |s|^{N+1} |\binom{-1/2}{N+1}| / (1-|s|)$.

```
      SUBROUTINE FSETZERO(v)
C..v(z)=0
      INCLUDE BIGN.STUFF
      DIMENSION v(1)
      DO 1 I=1,NERR
      v(I)=sZERO
    1 CONTINUE
      RETURN
      END




      SUBROUTINE FSETLIN(v,sC,sL)
C..v(z)=sC+z*sL
      INCLUDE BIGN.STUFF
      DIMENSION v(1)
      CALL FSETZERO(v)
      v(1)=sC
      v(2)=sL
      RETURN
      END




      SUBROUTINE FSETINV(v,s)
C..v(z)=1/(1+sz)
      INCLUDE BIGN.STUFF
      DIMENSION v(1)
      rT1=rMAXABS(s)
      IF(rT1.GE.rONE)WRITE(6,*)´error in FSETINV, radius=´,rT1
      v(1)=sONE
      DO 1 I=2,N1
      v(I)=sNEG(sPROD(v(I-1),s))
    1 CONTINUE
      CALL FCLEAN(v)
      s2=sABS(s)
      rTH=DIMAG(sQUOT(sPOWER(s2,N1),sDIFF(sONE,s2)))
      v(NERR)=error(rZERO,rTH)
      RETURN
      END
```

```
      SUBROUTINE FSETLOG(v,s)
C..v(z)=log(1+zs)
      INCLUDE BIGN.STUFF
      DIMENSION v(1)
      rT1=rMAXABS(s)
      IF(rT1.GE.rONE)WRITE(6,*)'error in FSETLOG, radius=',rT1
      s1=sNEG(s)
      s2=s
      v(1)=sZERO
      DO 1 I=1,N
      v(I+1)=sQUOT(s2,siCONST(I))
      s2=sPROD(s2,s1)
    1 CONTINUE
      CALL FCLEAN(v)
      s2=sABS(s)
      rTH=DIMAG(sQUOT(sQUOT(sPOWER(s2,N1),siCONST(N1)),sDIFF(sONE,s2)))
      v(NERR)=error(rZERO,rTH)
      RETURN
      END




      SUBROUTINE FSETMSQRT(v,s)
C..v(z)=(1+zs)**(-1/2)
      INCLUDE BIGN.STUFF
      DIMENSION v(1)
      rT1=rMAXABS(s)
      IF(rT1.GE.rONE)WRITE(6,*)'error in FSETMSQRT, radius=',rT1
      v(1)=sONE
      sF=sPROD(sNEG(sHALF),s)
      DO 1 I=2,N1
      v(I)=sF
      sI=siCONST(I)
      sF=sPROD(s,sPROD(sF,sQUOT(sDIFF(sHALF,sI),sI)))
    1 CONTINUE
      CALL FCLEAN(v)
      rTH=DIMAG(sQUOT(sABS(sF),sDIFF(sONE,sABS(s))))
      v(NERR)=error(rZERO,rTH)
      RETURN
      END
```

4.4.4 Functions

These are maps from B to S or R. The functions we consider are the norm, and evaluations.

We begin with $rL1NORM(\underline{v})$. (See Section 5 for $rNORM(\underline{v})$.)

Given $\underline{v} \in B$, we have for every $f \in A(\underline{v})$

$$|f|_1 \leq \sum_{i=0}^{N} |v_i| +_u v^{(G)} +_u v^{(H)} .$$

Given a scalar $s \in S$ and $\underline{v} \in B$, we have for every $f \in A(\underline{v})$ and every $z \in s$,

$$f(z) \in sVALUE(\underline{v}, s) ,$$

where $sVALUE$ is defined if $s \in [-1, 1]$ and is the interval

$$\sum_{i=0}^{N} v_i s^i +_s d_G +_s d_H$$

(all operations $S \times S \rightarrow S$) $d_G, d_H \in S$,

$$d_G \equiv [-1, 1] \cdot \{v^{(G)} *_u \max_r |s|\}$$

$$d_H \equiv [-1, 1] \cdot \{v^{(H)} *_u [\max_r |s|]^{N+1}\} .$$

Here, and below, powers are upward rounded.

Finally, given $s \in S$, $\underline{v} \in B$, and $j > 1$ we define $sDVALUE$ in such a way that for $s \subset [-1,1]$ and every $f \in A(\underline{v})$ and $z \in s$

$$\frac{d^j}{dz^j} f(z) \in sDVALUE(j,\underline{v},s) \quad .$$

One has the bounds

$$\sum_{k=0}^{N-j} (k+j)(k+j-1)\ldots(k+1) \cdot v_{k+j} \cdot s^k + d_{j,G} + d_{j,H}$$

(all operations $S \times S \to S$ (with natural injection of integers $\to S$)) with

$$d_{j,G} = \{v^{(G)} \cdot_u rSUP(\max_r |s|,1,j)\} \cdot [-1,1]$$

$$d_{j,H} = \{v^{(H)} \cdot_u rSUP(\max_r \tfrac{1}{3}|s|,N+1,j)\} \cdot [-1,1]$$

and

$$rSUP : R \times \mathbb{Z}_+ \times \mathbb{Z}_+ \to R$$

is an upper bound :

$$rSUP(x,k,j) \geq \sup_{i \geq k} (\frac{d}{dx})^j x^i \quad .$$

(rSUP will return an error message if x is too close to 1 and $j > 0$).

```
         REAL FUNCTION rL1NORM*8(v)
         ENTRY rNORM(v)
         INCLUDE BIGN.STUFF
         DIMENSION v(1)
         s=sSUM(sCONST(rG(v)),sCONST(rH(v)))
         DO 1 I=1,N1
         s=sSUM(s,sABS(v(N2-I)))
       1 CONTINUE
         r=rMAXABS(s)
         rL1NORM=r
         rNORM=r
         RETURN
         END
```

```
      REAL FUNCTION rSUP*8(r,K,J)
C..computes sup (d/dr)**J (r**I), for I>=K
C   r must be nonnegative
      INCLUDE BIGN.STUFF
      IF(J.LT.0 .OR. K.LT.J .OR. r.LT.rZERO
     *    .OR. (r.GT.0.99D+00 .AND. (r.GT.rONE .OR. J.NE.0)))
     *THEN
         WRITE(6,*)´incorrect arguments in rSUP,r,K,J´,r,K,J
         rSUP=rZERO
         RETURN
      ENDIF
      IF(J.EQ.0)THEN
         rSUP=DIMAG(sPOWER(DCMPLX(rZERO,r),K))
         RETURN
      ENDIF
C..the sup is attained for I<= J + J/log(1/r)-1
      NMAX=J-J/LOG(r)
      NMAX=MAXO(NMAX,K)
      rSUP=rZERO
C..rF=K*(K-1)*...*(K-J+1)*r**(K-J)
      rF=DIMAG(sPOWER(DCMPLX(rZERO,r),K-J))
      DO 1 I=1,J
      rF=rUP(rF*DFLOAT(K+1-I))
    1 CONTINUE
      DO 2 I=K,NMAX
      rSUP=rMAX2(rF,rSUP)
      rF=rUP(rUP(rUP(rF*DFLOAT(I+1)) *r) /DFLOAT(I-J+1))
    2 CONTINUE
      RETURN
      END




      COMPLEX FUNCTION sDVALUE*16(J1,v,s)
C..computes (d/dz)**J1 v (s)
C   should not be used with argument zero (use explicit formula!)
      INCLUDE BIGN.STUFF
      DIMENSION v(1)
      J=J1
      GOTO 11
      ENTRY sVALUE(v,s)
      J=0
   11 CONTINUE
      s1=sZERO
      DO 1 I=J+1,N1
      s2=sONE
      DO 2 K=1,J
      s2=sPROD(s2,siCONST(N2+J-I-K))
    2 CONTINUE
      s1=sSUM(sPROD(s,s1)
     *          ,sPROD(v(N2+J-I),s2))
    1 CONTINUE
      rS=rMAXABS(s)
      rT1=rUP(rSUP(rS,N1,J)*rH(v))
      rT2=rUP(rSUP(rS,MAXO(J,1),J)*rG(v))
      rT1=rUP(rT1+rT2)
      sDVALUE=sSUM(s1,DCMPLX(-rT1,rT1))
      sVALUE=sDVALUE
      RETURN
      END
```

4.4.5 <u>Maps in A</u>

These are functions which perform the sum, product,... of
elements in standard balls.

4.4.5.1. <u>Linear operations</u>

We first list some easy linear operations. If $\underline{v} \in B$ and
$f \in A(\underline{v})$ we need assignment (FEQUAL), negation (FNEG), multiplication and division by the argument (FSHFTR, FSHFTL).

We discuss only FSHFTL as an example and leave the analysis
of the other cases to a study of the printout.

If $\underline{v} \in B$ and $v_o = [0,0]$, then, <u>because the "general term"</u>
<u>of</u> $f \in A(\underline{v})$ <u>is</u> $\mathcal{O}(z)$ by construction we see that f itself is
$\mathcal{O}(z)$. Hence division by z is possible in this case. We want to
find $\underline{w} \in B$ such that if $f \in A(\underline{v})$ then $f(z)/z \in A(\underline{w})$. There are
several possible ways to define $\underline{w}$ and we are forced to choose
one. E.g. the "higher order term", when divided by z, is no longer
guaranteed to be $\mathcal{O}(z^{N+1})$. We can <u>either</u> consider it as part of
the "general term" of $\underline{w}$ <u>or</u> assume the worst and account for both
the possibility that it is exactly a monomial of degree N or
that it is a higher order term or a combination of both. We choose
the second possibility because we want to keep as much information
about the order of a function near zero. We also have to make sure
that the "general term" of $\underline{w}$ is $\mathcal{O}(z)$ by defining

$$w_o = v_1 +_s [-v^{(G)}, v^{(G)}] .$$

Discussions as the one just given are necessary in many places, but will not be repeated.

```
          SUBROUTINE FEQUAL(v,vRES)
C..vRES(z)=v(z)
          INCLUDE BIGN.STUFF
          DIMENSION v(1),vRES(1)
          DO 1 I=1,NERR
          vRES(I)=v(I)
        1 CONTINUE
          RETURN
          END


          SUBROUTINE FNEG(v,vRES)
C..vRES(z)=-v(z)
          INCLUDE BIGN.STUFF
          DIMENSION v(1),vRES(1)
          DO 1 I=1,N1
          vRES(I)=sNEG(v(I))
        1 CONTINUE
          CALL FCLEAN(vRES)
          vRES(NERR)=v(NERR)
          RETURN
          END


          SUBROUTINE FSHFTL(v,vRES)
C..vRES(z)=v(z)/z
          INCLUDE BIGN.STUFF
          DIMENSION v(1),vRES(1)
          IF(rMAXABS(v(1)).NE.rZERO)THEN
             WRITE(6,*)'error in FSHFTL, v(1) is not zero',v(1)
          ENDIF
          DO 1 I=1,N
          vRES(I)=v(I+1)
        1 CONTINUE
          CALL FCLEAN(vRES)
          r=rG(v)
          vRES(1)=sSUM(vRES(1),DCMPLX(-r,r))
          r=rH(v)
          vRES(N1)=DCMPLX(-r,r)
          vRES(NERR)=v(NERR)
          RETURN
          END


          SUBROUTINE FSHFTR(v,vRES)
C..vRES(z)=v(z)*z
          INCLUDE BIGN.STUFF
          DIMENSION v(1),vRES(1)
          sT1=v(N1)
          DO 1 I=1,N
          vRES(N2-I)=v(N1-I)
        1 CONTINUE
          CALL FCLEAN(vRES)
          vRES(1)=sZERO
          vRES(NERR)=eSUM(v(NERR),error(rZERO,rMAXABS(sT1)))
          RETURN
          END
```

The next operations are multiplication with a scalar
$A(\underline{w}) \ni x \cdot f$, $x \in s$, $f \in A(\underline{v})$, (FsMULT); $A(\underline{w}) \ni x \cdot f_1 + f_2$, $x \in s$, $f_1 \in A(\underline{v}_1)$, $f_2 \in A(\underline{v}_2)$ (FsMULADD). Again, the estimates are obvious.

```
      SUBROUTINE FsMULT(v,s,vRES)
C..vRES(z)=s*v(z)
      INCLUDE BIGN.STUFF
      DIMENSION v(1),vRES(1)
      DO 1 I=1,N1
      vRES(I)=sPROD(s,v(I))
    1 CONTINUE
      CALL FCLEAN(vRES)
      sTG=sPROD(s,sCONST(rG(v)))
      sTH=sPROD(s,sCONST(rH(v)))
      vRES(NERR)=error(rMAXABS(sTG),rMAXABS(sTH))
      RETURN
      END


      SUBROUTINE FsMULADD(v,s,vRES)
C..vRES(z)=vRES(z)+s*v(z)
      INCLUDE BIGN.STUFF
      DIMENSION v(1),vRES(1)
      DO 1 I=1,N1
      vRES(I)=sSUM(vRES(I),sPROD(s,v(I)))
    1 CONTINUE
      rTG=rMAXABS(sPROD(s,sCONST(rG(v))))
      rTH=rMAXABS(sPROD(s,sCONST(rH(v))))
      vRES(NERR)=eSUM(error(rTG,rTH),vRES(NERR))
      RETURN
      END
```

We next include addition and subtraction.

```
      SUBROUTINE FADD(v,vSUM)
C..vSUM(z)=vSUM(z)+v(z)
      INCLUDE BIGN.STUFF
      DIMENSION v(1),vSUM(1)
      DO 1 I=1,N1
      vSUM(I)=sSUM(vSUM(I),v(I))
    1 CONTINUE
      vSUM(NERR)=eSUM(v(NERR),vSUM(NERR))
      RETURN
      END


      SUBROUTINE FMINUS(v,vDIFF)
C..vDIFF(z)=vDIFF(z)-v(z)
      INCLUDE BIGN.STUFF
      DIMENSION v(1),vDIFF(1)
      DO 1 I=1,N1
      vDIFF(I)=sDIFF(vDIFF(I),v(I))
    1 CONTINUE
      vDIFF(NERR)=eSUM(v(NERR),vDIFF(NERR))
      RETURN
      END
```

4.4.5.2. Multiplication

The multiplication, FMULT, is straightforward to implement. Given $\underline{v}, \underline{w} \in B$, we want to find an $\underline{u} \in B$ such that for every $f \in A(\underline{v})$ and every $g \in A(\underline{w})$, one has $f \cdot g \in A(\underline{u})$. The product $f \cdot g$ can be conveniently estimated by considering the polynomial, general and higher order parts of f and g (i.e. of $\underline{v}$ and $\underline{w}$). E.g., the product of the polynomial part yields a polynomial plus a higher order contribution to $\underline{w}$. Other cases are similar, see the program and Eq. (4.3) below. We also use FMULTADD, which adds a product to an existing vector.

```
      SUBROUTINE FMULT(vA,vB,vRES)
C..vRES(z)=vA(z)*vB(z)
      INCLUDE BIGN.STUFF
      DIMENSION vA(1),vB(1),vRES(1)
      DIMENSION xT1(NDEG)
      CALL FSETZERO(xT1)
      DO 1 K=0,N
      s1=sZERO
      DO 2 I=0,K
      s1=sSUM(s1,sPROD(vA(I+1),vB(K-I+1)))
    2 CONTINUE
      xT1(K+1)=s1
    1 CONTINUE
      sH=sZERO
      DO 3 K=N+1,N+N
      s1=sZERO
      DO 4 I=K-N,N
      s1=sSUM(s1,sPROD(vA(I+1),vB(K-I+1)))
    4 CONTINUE
      sH=sSUM(sH,sABS(s1))
    3 CONTINUE
      sAG=sCONST(rG(vA))
      sAH=sCONST(rH(vA))
      sBG=sCONST(rG(vB))
      sBH=sCONST(rH(vB))
      s1=sZERO
      s2=sZERO
      DO 5 I=1,N
      s1=sSUM(s1,sABS(vA(N1-I)))
      s2=sSUM(s2,sABS(vB(N1-I)))
    5 CONTINUE
      sH=sSUM(sH,sSUM(sPROD(sABS(vA(N1)),sBG),sSUM(sPROD(sABS(vB(N1))
     *    ,sAG),sSUM(sPROD(sAG,sBH),sSUM(sPROD(sAH,sBG),sSUM(sPROD
     *    (sAH,sBH),sSUM(sPROD(sAH,sSUM(s2,sABS(vB(N1)))),sPROD(sBH
     *    ,sSUM(s1,sABS(vA(N1))))))))))))
      sG=sSUM(sPROD(sAG,sBG),sSUM(sPROD(s1,sBG),sPROD(s2,sAG)))
      CALL FxeTOv(xT1,error(rMAXABS(sG),rMAXABS(sH)),vRES)
      RETURN
      END
```

We have

$$u^{(H)} = |\text{higher order part of polynomial } (v \cdot w)|_1$$

$$+ |v_N| \cdot w^{(G)} + |w_N| \cdot v^{(G)}$$

$$+ v^{(G)} \cdot w^{(H)} + w^{(G)} \cdot v^{(H)} \tag{4.3}$$

$$+ v^{(H)} \cdot w^{(H)} + v^{(H)} \cdot \sum_{i=0}^{N} |w_i| + w^{(H)} \cdot \sum_{i=0}^{N} |v_i|$$

$$u^{(G)} = v^{(G)} \cdot w^{(G)} + \sum_{i=0}^{N-1} |v_i| w^{(G)} + \sum_{i=0}^{N-1} |w_i| v^{(G)} \, .$$

```
        SUBROUTINE FMULTADD(vA,vB,vRES)
C..vRES(z)=vRES(z)+vA(z)*vB(z)
        INCLUDE BIGN.STUFF
        DIMENSION vA(1),vB(1),vRES(1)
        DIMENSION vT1(NDEG)
        CALL FMULT(vA,vB,vT1)
        CALL FADD(vT1,vRES)
        RETURN
        END
```

4.4.5.3. Composition, and related operations

We describe now the operations $f \circ g$, $f(s.)$, $f'(s.)$, $f' \circ g$.
We shall treat in great detail the first operation, for which we
need relatively detailed estimates. The remaining cases are then
simple variants and can be easily read off the program.

At the basis of all estimates for composition in A is the
observation that if $|f|_1 < \infty$ and $|g|_1 \leq 1$ then the range of
g, i.e. $\{g(z)|z \in \bar{D}\}$ is contained in $\bar{D}$ (and hence $f \circ g$ is

defined) and one has the bound $|f \circ g|_1 \leq |f|_1$. This follows at once from the identity

$$f \circ g(z) = \sum_{j=0}^{\infty} f_j \, g(z)^j \, ,$$

$$|f \circ g|_1 \leq \sum_{j=0}^{\infty} |f_j| |g|_1^j \, .$$

This also implies that if $f(z) = O(z^k)$ then $|f \circ g|_1 \leq |f|_1 |g|^k$.

We use these estimates when $|g|_1 \leq 1$. Now it may happen that $g(\bar{D}) \subset \bar{D}$ but $|g|_1 > 1$, and in such cases it is still possible to bound $|f \circ g|_1$. Our algorithm (suggested by Lanford and Llave [51]) works provided $|g^k|_1 < 1$ for some $k \leq N + 1$. In that case we use the bound

$$|f \circ g|_1 \leq \sum_{j=0}^{k} |f_j| |g^j|_1 + \sum_{j=k+1}^{\infty} |f_j| |g^k|_1^{[j/k]} \cdot \sup_{k > \ell \geq 0} |g^\ell|_1$$

$$\leq \sum_{j=0}^{k} |f_j| |g^j|_1 + \sup_{k > \ell \geq 0} |g^\ell|_1 \sum_{j=k+1}^{\infty} |f_j| |g^k|_1 \, .$$

In fact, we apply the inequality with $k = N + 1$, i.e. we require $|g^{N+1}|_1 < 1$.

In the case when $|g|_1 \leq 1$, we distinguish the subcases $g(0) \neq 0$ and $g(0) = 0$. In the first case, the errors in $w = f \circ g$ are bounded by decomposing $g = g(0) + (g - g(0)) = g_0 + g_1$ and we have

$$w^{(G)} = f^{(G)} \cdot |g_1|_1 + f^{(H)} \{ |g|_1^{N+1} - |g_1|_1^{N+1} - |g_0|^{N+1} \}$$

$$+ \text{ contribution form } \sum_{i=0}^{N} f_i g^i \, .$$

$$w^{(H)} = f^{(H)} |g_1|_1^{N+1} + \text{contribution from } \sum_{i=0}^{N} f_i g^i$$

and we define $w_o = \{ |g_o| f^{(G)} + |g_o|^{N+1} f^{(H)} \} \cdot [-1,1] + \text{contribu-}$
tion from $\sum_{i=0}^{N} f_i g^i$.

If $g(0) = 0$, then we use the bound

$$w^{(G)} = f^{(G)} |g|_1 + \text{contribution from } \sum_{i=0}^{N} f_i g^i \quad,$$

$$w^{(H)} = f^{(H)} |g|_1^{N+1} + \text{contribution from } \sum_{i=0}^{N} f_i g^i \quad.$$

```
      SUBROUTINE FCOMP(v,vZ,vRES)
C..vRES(z)=v(vZ(z))
      INCLUDE BIGN.STUFF
      DIMENSION v(1),vZ(1),vRES(1)
      DIMENSION vT1(NDEG),vT2(NDEG),vT3(NDEG)
      DIMENSION vSAVE(6,NDEG)
C..compute substitution into polynomial
      CALL FSETZERO(vT1)
      vT1(1)=v(N1)
      DO 2 I=1,N
      CALL FMULT(vT1,vZ,vT1)
      vT1(1)=sSUM(vT1(1),v(N1-I))
    2 CONTINUE
      IF(rL1NORM(vZ).GT.rONE)GOTO 1
C..error estimate in case of norm <=1
      sTVG=sCONST(rG(v))
      sTVH=sCONST(rH(v))
      sTN1=sCONST(rL1NORM(vZ))
      IF(rMAXABS(vZ(1)).EQ.rZERO)GOTO 3
C..generic case: vZ(z) has constant term
      s1=vZ(1)
      vZ(1)=sZERO
      sTN2=sCONST(rL1NORM(vZ))
      vZ(1)=s1
      s1=sCONST(rMAXABS(s1))
      s1N1=sPOWER(s1,N1)
      rT0=rMAXABS(sSUM(sPROD(s1,sTVG),sPROD(s1N1,sTVH)))
      vT1(1)=sSUM(vT1(1),DCMPLX(-rT0,rT0))
      sTN2N1=sPOWER(sTN2,N1)
      s2=sCONST(rMAXABS(sDIFF(sPOWER(sTN1,N1),sSUM(sTN2N1,s1N1))))
      rTG=rMAXABS(sSUM(sPROD(sTN2,sTVG),sPROD(s2,sTVH)))
      rTH=rMAXABS(sPROD(sTN2N1,sTVH))
      vT1(NERR)=eSUM(vT1(NERR),error(rTG,rTH))
      GOTO 4
    3 CONTINUE
```

```
C..case when vZ(0)=0
      rTH=rMAXABS(sPROD(sPOWER(sTN1,N1),sTVH))
      vT1(NERR)=eSUM(vT1(NERR),error(rMAXABS(sPROD(sTN1,sTVG)),rTH))
    4 CONTINUE
      CALL FEQUAL(vT1,vRES)
      RETURN
    1 CONTINUE
C..error estimate in case norm >1
C  prepare table of powers of vZ
C  vSAVE(i,.)=vZ**(2**i)
      CALL FMULT(vZ,vZ,vT2)
      J=2
      DO 10 I=1,6
      DO 11 K=1,NERR
      vSAVE(I,K)=vT2(K)
   11 CONTINUE
      J=J+J
      IF(J.GT.N1)GOTO 12
      CALL FMULT(vT2,vT2,vT2)
   10 CONTINUE
   12 CONTINUE
C..begin determining maxima
C  r=sup{rL1NORM(vZ**k) | k >=0 }
C  r2=sup{rL1NORM(vZ**k - vZ**k(z=0)) | k >=0 }
C  rN1=rL1NORM(vZ**(N+1))
      r=rMAX2(rONE,rL1NORM(vZ))
      sSAVE=vZ(1)
      vZ(1)=sZERO
      r2=rL1NORM(vZ)
      vZ(1)=sSAVE
      DO 13 I=3,N2,2
      K=(I-1)/2
      J=1
      CALL FSETZERO(vT2)
      vT2(1)=sONE
C..loop to produce powers
   14 CONTINUE
      KH=K/2
      IF(KH+KH.NE.K)THEN
        DO 15 L=1,NERR
        vT3(L)=vSAVE(J,L)
   15   CONTINUE
        CALL FMULT(vT2,vT3,vT2)
        K=K-1
      ENDIF
      IF(K.EQ.0)GOTO 16
      K=KH
      J=J+1
      GOTO 14
   16 CONTINUE
C..take max
      CALL FMULT(vT2,vZ,vT3)
      r=rMAX3(r,rL1NORM(vT2),rL1NORM(vT3))
      sSAVE=vT2(1)
      vT2(1)=sZERO
      r2=rMAX2(r2,rL1NORM(vT2))
      vT2(1)=sSAVE
      sSAVE=vT3(1)
      vT3(1)=sZERO
      r2=rMAX2(r2,rL1NORM(vT3))
      vT3(1)=sSAVE
   13 CONTINUE
```

```
      NH=N/2
      IF(NH+NH.EQ.N)THEN
        rN1=rL1NORM(vT3)
      ELSE
        rN1=rL1NORM(vT2)
      ENDIF
C..end determining maxima
C..estimate error
      IF(rN1.GT.rONE)THEN
        WRITE(6,*)´error in FCOMP, rN1=´,rN1
      ENDIF
      sTVG=sCONST(rG(v))
      sTVH=sCONST(rH(v))
      sT1=sPOWER(sABS(vZ(1)),N1)
      rT0=rMAXABS(sSUM(sPROD(sABS(vZ(1)),sTVG),
     *                 sPROD(sT1,sTVH)))
      vT1(1)=sSUM(vT1(1),DCMPLX(-rT0,rT0))
      rT0=rMAXABS(sSUM(sPROD(sCONST(r2),sTVG),
     *                 sPROD(sCONST(rUP(r*rN1)),sTVH)))
      vT1(NERR)=eSUM(vT1(NERR),error(rT0,rZERO))
      CALL FEQUAL(vT1,vRES)
      RETURN
      END
```

We next present FsDILAT, where the error bounds should be obvious.

```
      SUBROUTINE FsDILAT(v,s,vRES)
C..vRES(z)=v(s*z)
      INCLUDE BIGN.STUFF
      COMMON/SWITCH/ITEST
      DIMENSION v(1),vRES(1)
      IF(ITEST.EQ.1.AND.rMAXABS(s).GT.rONE)THEN
        rT1=rMAXABS(s)
        WRITE(6,*)´error in FsDILAT, rMAXABS(s)=´,rT1
      ENDIF
      vRES(1)=v(1)
      s1=s
      DO 1 I=2,N1
      vRES(I)=sPROD(v(I),s1)
      s1=sPROD(s,s1)
    1 CONTINUE
      rTG=rMAXABS(sPROD(s,sCONST(rG(v))))
      rTH=rMAXABS(sPROD(s1,sCONST(rH(v))))
      CALL FCLEAN(vRES)
      vRES(NERR)=error(rTG,rTH)
      RETURN
      END
```

It is now easy to understand from the program the definitions
for f'(sz), and f'∘g. The subroutines CvTOxe, CDERIVATE act
on formal power series (of degree N), see below.

```
      SUBROUTINE FDZsDILAT(v,s,vRES)
C..vRES(z)=((d/dz)v)(s*z)
      INCLUDE BIGN.STUFF
      COMMON/SWITCH/ITEST
      DIMENSION v(1),vRES(1)
      DIMENSION x(NDEG)
      IF(ITEST.EQ.1.AND.rMAXABS(s).GT.0.99D+00)THEN
        r1=rMAXABS(s)
        WRITE(6,*)´error in FDZsDILAT,rMAXABS(s)=´,r1
      ENDIF
      CALL CvTOxe(v,x,e)
      CALL CDERIVATE(x,x)
      vRES(1)=x(1)
      s1=s
      DO 1 I=2,N1
      vRES(I)=sPROD(x(I),s1)
      s1=sPROD(s,s1)
    1 CONTINUE
      CALL FCLEAN(vRES)
      rTG=rG(v)
      rTH=rH(v)
      vRES(1)=sSUM(vRES(1),DCMPLX(-rTG,rTG))
      r=rMAXABS(sPROD(sPROD(siCONST(N1),sABS(s1)),sCONST(rTH)))
      vRES(N1)=sSUM(vRES(N1),DCMPLX(-r,r))
      rS=rMAXABS(s)
      sT1=sPROD(sCONST(rSUP(rS,2,1)),sCONST(rTG))
      sT2=sPROD(sCONST(rSUP(rS,N2,1)),sCONST(rTH))
      vRES(NERR)=error(rMAXABS(sT1),rMAXABS(sT2))
      RETURN
      END
```

We compute $f'\circ g$ as $f'(s.)\circ(\frac{1}{s}g)$ with $s = \min(0.935, |g|_1 \cdot 1.0001)$

```
      SUBROUTINE FDZCOMP(v,vZ,vRES)
C..vRES(z)=(d/dz vZ) (v(z))
      INCLUDE BIGN.STUFF
      DIMENSION v(1),vZ(1),vRES(1)
      DIMENSION vT1(NDEG)
      RR=.935D+00
      s=sCONST(rMIN2(RR,rL1NORM(vZ)*1.0001D+00))
      CALL FsMULT(vZ,sINV(s),vT1)
      CALL FDZsDILAT(v,s,vRES)
      CALL FCOMP(vRES,vT1,vRES)
      RETURN
      END
```

4.4.5.4. Inverses, Exponentials, Logarithm, Roots

These are straightforward estimates, and we have indicated the estimates in the programs themselves.

```
      SUBROUTINE FINVERS(v,vRES)
C..vRES(z)=1/v(z)
      INCLUDE BIGN.STUFF
      LOGICAL logZERO
      DIMENSION v(1),vRES(1)
      DIMENSION x(NDEG),v1(NDEG),v2(NDEG)
      IF(logZERO(v(1)))THEN
        WRITE(6,*)'error in FINVERS, v(1) might be zero'
      ENDIF
      rT1=rMAXABS(sQUOT(sCONST(rL1NORM(v)),v(1)))
      IF(rT1.GT.1.99D+00)THEN
        WRITE(6,*)'error in FINVERS, 1/v does not exist on this domain'
      ENDIF
C..upper bound on norm of (v-v(0))/v(0)
      s=sCONST(rT1-.99D+00)
      CALL FSETINV(v1,s)
      CALL FsMULT(v,sINV(sPROD(s,v(1))),v2)
      v2(1)=sZERO
      CALL FCOMP(v1,v2,v1)
      CALL FsMULT(v1,sINV(v(1)),v1)
C..recompute power series part
      CALL CvTOxe(v,x,e)
      CALL CINVERS(x,x)
      DO 1 I=2,N1
      vRES(I)=sINTER(v1(I),x(I))
    1 CONTINUE
      vRES(1)=v1(1)
      CALL FCLEAN(vRES)
      vRES(NERR)=v1(NERR)
      RETURN
      END
```

```
      SUBROUTINE FMSQRT(v,vRES)
C..vRES(z)=(v(z))**(-1/2)
      INCLUDE BIGN.STUFF
      DIMENSION v(1),vRES(1)
      DIMENSION v1(NDEG),v2(NDEG)
      IF(DREAL(v(1)).LE.rZERO)THEN
        WRITE(6,*)´error in FMSQRT, v(1) might be negative or zero´
      ENDIF
      rT1=rMAXABS(sQUOT(sCONST(rL1NORM(v)),v(1)))
      IF(rT1.GT.1.99D+00)THEN
        WRITE(6,*)´error in FMSQRT, (V)**(-1/2) does not exist ´
      ENDIF
C..upper bound on norm of (v-v(0))/v(0)
      s=sCONST(rT1-.99D+00)
      CALL FSETMSQRT(v1,s)
      CALL FsMULT(v,sINV(sPROD(s,v(1))),v2)
      v2(1)=sZERO
      CALL FCOMP(v1,v2,v1)
      CALL FsMULT(v1,sINV(sSQRT(v(1))),vRES)
      RETURN
      END




      SUBROUTINE FEXP(v,vRES)
C..vRES(z)=exp(v(z)); v(0) has to be zero
      INCLUDE BIGN.STUFF
      DIMENSION v(1),vRES(1)
      DIMENSION vT1(NDEG),vT2(NDEG)
      IF(rMAXABS(v(1)).NE.rZERO)THEN
        WRITE(6,*)´error in FEXP, substituend not of desired form´
      ENDIF
C..we compute vRES(z)=exp(norm(v)*(v(z)/norm(v)))
C   the error is then bounded by norm**N1/(N1!*(1-norm/N2))
      s1=saCONST(rL1NORM(v)*1.0001D+00)
      IF(rMAXABS(s1).GE.DFLOAT(N2))THEN
        WRITE(6,*)´error in FEXP, degree not high enough´
      ENDIF
      CALL FSETZERO(vT1)
      vT1(1)=sONE
      DO 1 I=1,N
      vT1(I+1)=sQUOT(sPROD(vT1(I),s1),siCONST(I))
    1 CONTINUE
      s2=sABS(vT1(N1))
      CALL FsMULT(v,sINV(s1),vT2)
      CALL FCOMP(vT1,vT2,vRES)
      s3=sCONST(rH(vRES))
      I=N1
    2 CONTINUE
      s2=sABS(sQUOT(sPROD(s2,s1),siCONST(I)))
      s3=sABS(sSUM(s3,s2))
      I=I+1
      IF(rMAXABS(s2).GT.1.D-25)GOTO 2
      s3=sSUM(s3,sQUOT(sPROD(s2,s1),siCONST(I))
     *   ,sDIFF(sONE,sQUOT(s1,siCONST(I+1)))))
      vRES(NERR)=error(rG(vRES),rMAXABS(s3))
      RETURN
      END
```

```
      SUBROUTINE FLOG(v,vRES)
C..vRES(z)=log(v(z)); v(0) has to be 1
      INCLUDE BIGN.STUFF
      DIMENSION v(1),vRES(1)
      DIMENSION x(NDEG),v1(NDEG),v2(NDEG)
      IF(DREAL(v(1)).NE.rONE.OR.DIMAG(v(1)).NE.rONE)THEN
        WRITE(6,*)'error in FLOG, v(1) is not one'
        STOP
      ENDIF
      rT1=rMAXABS(sQUOT(sCONST(rL1NORM(v)),v(1)))
      IF(rT1.GT.1.99D+00)THEN
        WRITE(6,*)'error in FLOG, log(v) does not exist on this domain'
      ENDIF
C..upper bound on norm of (v-v(0))/v(0)
      s=sCONST(rT1-.99D+00)
      CALL FSETLOG(v1,s)
      CALL FsMULT(v,sINV(sPROD(s,v(1))),v2)
      v2(1)=sZERO
      CALL FCOMP(v1,v2,v1)
C..recompute power series part
      CALL CvTOxe(v,x,e)
      CALL CLOG(x,x)
      DO 1 I=2,N1
      vRES(I)=sINTER(v1(I),x(I))
    1 CONTINUE
      vRES(1)=sZERO
      CALL FCLEAN(vRES)
      vRES(NERR)=v1(NERR)
      RETURN
      END
```

4.5. Appendices

Here, we list some trivial subroutines which complete the picture, but which need no special explanations.

4.5.1 Operations acting on formal power series

4.5.1.1. Make power series

```
      SUBROUTINE CvTOxe(v,x,e)
C..split vector into formal power series and error
      INCLUDE BIGN.STUFF
      DIMENSION v(1),x(1)
      DO 1 I=1,N1
      x(I)=v(I)
    1 CONTINUE
      e=v(NERR)
      CALL CCLEAN(x)
      RETURN
      END


      SUBROUTINE CSETLOG(x)
C..x(z)=-log(1-z)
      INCLUDE BIGN.STUFF
      DIMENSION x(1)
      x(1)=sZERO
      DO 1 I=1,N
      x(I+1)=sINV(siCONST(I))
    1 CONTINUE
      CALL CCLEAN(x)
      RETURN
      END


      SUBROUTINE CSETINV(x)
C..x(z)=1/(1+z)
      INCLUDE BIGN.STUFF
      DIMENSION x(1)
      x(1)=sONE
      DO 1 I=2,N1
      x(I)=sNEG(x(I-1))
    1 CONTINUE
      CALL CCLEAN(x)
      RETURN
      END
```

4.5.1.2. Basic arithmetic operations. Structure : x1 = C(x2) or x = C(x)

```
      SUBROUTINE CSHFTL(x,xRES)
C..xRES(z)=x(z)/z
      INCLUDE BIGN.STUFF
      DIMENSION x(1),xRES(1)
      IF(rMAXABS(x(1)).GT.rZERO)THEN
         WRITE(6,*)'error in CSHFTL, x(1) is not zero',x(1)
      ENDIF
      DO 1 I=1,N
      xRES(I)=x(I+1)
    1 CONTINUE
      xRES(N1)=sZERO
      CALL CCLEAN(xRES)
      RETURN
      END


      SUBROUTINE CSHFTR(x,xRES)
C..xRES(z)=x(z)*z
      INCLUDE BIGN.STUFF
      DIMENSION x(1),xRES(1)
      DO 1 I=1,N
      xRES(N2-I)=x(N1-I)
    1 CONTINUE
      xRES(1)=sZERO
      CALL CCLEAN(xRES)
      RETURN
      END


      SUBROUTINE CINVERS(x,xRES)
C..xRES(z)=1/x(z)
      INCLUDE BIGN.STUFF
      LOGICAL logZERO
      DIMENSION x(1),xRES(1)
      DIMENSION xT1(NDEG)
      IF(logZERO(x(1)))THEN
         WRITE(6,*)'error in CINV, x(1) might be zero'
C        STOP
      ENDIF
      CALL FSETZERO(xT1)
      s1=sINV(x(1))
      xT1(1)=s1
      DO 2 I=2,N1
      s=sZERO
      DO 3 J=2,I
      s=sSUM(s,sPROD(x(J),xT1(I+1-J)))
    3 CONTINUE
      xT1(I)=sNEG(sPROD(s,s1))
    2 CONTINUE
      CALL FEQUAL(xT1,xRES)
      RETURN
      END
```

```
      SUBROUTINE CLOG(x,xRES)
C..xRES(z)=log(x(z))
      INCLUDE BIGN.STUFF
      DIMENSION x(1),xRES(1)
      DIMENSION xEXP(NDEG),xT1(NDEG),xT2(NDEG)
      NSAVE=N
      xEXP(1)=sONE
      DO 1 I=1,N
      xEXP(I+1)=sQUOT(xEXP(I),siCONST(I))
    1 CONTINUE
      CALL CCLEAN(xEXP)
      CALL FSETZERO(xT1)
      DO 2 I=2,NSAVE+1
      CALL DEG(I-1)
      CALL CCOMP(xEXP,xT1,xT2)
      xT1(I)=sDIFF(x(I),xT2(I))
    2 CONTINUE
      CALL FEQUAL(xT1,xRES)
      RETURN
      END


      SUBROUTINE CDERIVATE(x,xRES)
C..xRES(z)=(d/dz)x(z)
      INCLUDE BIGN.STUFF
      DIMENSION x(1),xRES(1)
      DO 1 I=1,N
      xRES(I)=sPROD(x(I+1),siCONST(I))
    1 CONTINUE
      xRES(N1)=sZERO
      CALL CCLEAN(xRES)
      RETURN
      END


      SUBROUTINE CCLEAN(x)
C..set unused space to zero
      INCLUDE BIGN.STUFF
      DIMENSION x(1)
      DO 1 I=N2,NERR
      x(I)=sZERO
    1 CONTINUE
      RETURN
      END
```

4.5.1.3. Arithmetic operations. Structure : x1 = C(x2,x3)

```
      SUBROUTINE CMULT(xA,xB,xRES)
C..xRES(z)=xA(z)*xB(z)
      INCLUDE BIGN.STUFF
      DIMENSION xA(1),xB(1),xRES(1)
      DO 1 II=1,N1
      K=N2-II
      s1=sZERO
      DO 2 J=1,K
      s1=sSUM(s1,sPROD(xA(K-J+1),xB(J)))
    2 CONTINUE
      xRES(K)=s1
    1 CONTINUE
      CALL CCLEAN(xRES)
      RETURN
      END


      SUBROUTINE CCOMP(x,xZ,xRES)
C..xRES(z)=x(xZ(z))
      INCLUDE BIGN.STUFF
      DIMENSION x(1),xZ(1),xRES(1)
      DIMENSION xT1(NDEG)
      CALL FSETZERO(xT1)
      xT1(1)=x(N1)
      DO 1 I=1,N
      CALL CMULT(xT1,xZ,xT1)
      xT1(1)=sSUM(xT1(1),x(N1-I))
    1 CONTINUE
      CALL FEQUAL(xT1,xRES)
      RETURN
      END
```

4.5.2 (Computer-) arithmetic operations

```
      SUBROUTINE AvTOp(v,p)
      ENTRY AxTOp(v,p)
      INCLUDE BIGN.STUFF
      DIMENSION v(1),p(1)
      DO 1 I=1,N1
      p(I)=aCENT(v(I))
    1 CONTINUE
      RETURN
      END


      REAL FUNCTION aDXVALUE*8(p,a)
      INCLUDE BIGN.STUFF
      DIMENSION p(1)
      aN=N
      aSUM=p(N1)*aN
      DO 1 I=2,N
      aN1I=N1-I
      aSUM=aSUM*a+p(N2-I)*aN1I
    1 CONTINUE
      aDXVALUE=aSUM
      RETURN
      END


      REAL FUNCTION aDXXVALUE*8(p,a)
      INCLUDE BIGN.STUFF
      DIMENSION p(1)
      aN=N
      aSUM=p(N1)*aN*(aN-rONE)
      DO 1 I=2,N-1
      aN1I=N1-I
      aSUM=aSUM*a+p(N2-I)*aN1I*(aN1I-rONE)
    1 CONTINUE
      aDXXVALUE=aSUM
      RETURN
      END
```

5. The equations at $N = \infty$

5.1. Reformulation of the problem

In the introduction to this paper we have motivated that we want to solve the system of equations (1.7)-(1.10). Furthermore, we explained that we want to achieve this by applying the contraction mapping principle to a suitable operator K_{∞}. However, in order to give mathematical sense to K_{∞} we first need to reformulate the problem in terms of normal forms as described in Section 2.2.

First of all, we fix the origin of our coordinate system at the fixed point of the function g of (1.7)-(1.10), i.e. we rewrite g as

$$g(z) = G(-x_{\infty} + z) + x_{\infty} ,$$

where $x_{\infty} \simeq 0.391$ is a fixed normalization constant which fixes the length scale in the complex plane. (See Theorem 1.1 for the exact value of x_{∞}.) The function G has a Taylor expansion of the form $G(z) = -z + O(z^2)$, provided g solves (1.7)-(1.10). We define now, as described in Section 2.2, the function $G_o(z) = (T_A)^{-1} \circ G \circ T_A(z)$, where $T_A(z) = z/(1 + Az)$ and $A = -\frac{1}{4}G''(0)$.

We change now perspective, and view A and G_o as the basic quantities. The problem to solve (1.7)-(1.10) can then be formu-

lated as a fixed point problem for an operator K_o ,
$\{A,G_o\} \to \{\tilde{A},\tilde{G}_o\}$ which is defined through the following sequence of operations.

i) given $\{A,G_o(z)\}$, $(G_o(z) = -z + 0(z^3))$ define

$$\Gamma(z) = G_o(G_o(z)) , \qquad (5.1.1)$$

ii) solve the equation

$$F_o(\Gamma(z)) - F_o(z) = -2 , \qquad (5.1.2)$$

iii) define $h(z) = F_o(T_{-A}(-x_{oo} + z)) - F_o(T_{-A}(-x_{oo}))$, $\qquad$ (5.1.3)

iv) solve the following equation for τ :

$$1 - \log\tau \cdot \tau \cdot h'(\tau x_{oo}) = 0 , \qquad (5.1.4)$$

v) set $\alpha = x_{oo} \cdot \exp(\log\tau \cdot h(\tau \cdot x_{oo}))$, $\qquad$ (5.1.5)

vi) define $\tilde{G}(z) = \alpha \cdot \exp(-\log\tau \cdot h(\tau \cdot x_{oo} + \tau z)) - x_{oo}$, $\qquad$ (5.1.6)

vii) determine $\tilde{A} = -\frac{1}{4}(\frac{1}{x_{oo}} + x_{oo}(-\log\tau)\tau^2 \cdot h''(\tau \cdot x_{oo}))$, $\qquad$ (5.1.7)

viii) define $\tilde{G}_o(z) = T_{-\tilde{A}} \circ \tilde{G} \circ T_{\tilde{A}}(z)$, $\qquad$ (5.1.8)

ix) define $K_o(\{A,G_o(z)\}) = \{\tilde{A},\tilde{G}_o(z)\}$. $\qquad$ (5.1.9)

Remark : It is vi) which necessitates a computer and ii) which is delicate and causes the singularity of the problem.

Lemma 5.1.1. If K_o is defined and has a fixed point, then (1.7)-(1.10) has a solution.

This is obvious from the definition of K_o.

The main problem with K_o is the constructive estimation of the solution of ii). The only other equation to be solved is iv). This is done with a constructive variant of the Newton algorithm (see Section 4 and below for details). All other operations are trivial mathematics.

We want to use the contraction mapping principle to prove the existence of a fixed point for K_o. Thus, as we have already described in the case of the operator K_{oo}, we need to find a choice of $\{A, G_o\}$, for which, in a suitable norm

$$\| K_o(\{A, \hat{G}_o\}) - \{A, \hat{G}_o\} \| < \varepsilon .$$

Then we bound the tangent map DK_o on a ball of radius β around $\{A, \hat{G}_o\}$, i.e. we show that

$$\| DK_{o\{A,G_o\}} \| \leq \rho < 1 ,$$

for

$$\| \{A, G_o\} - \{A, \hat{G}_o\} \| < \beta .$$

The detailed description and implementation of K_o will be given later. To orient the reader, we also describe the action of the

linear operator DK_o at $\{A,G_o\}$, acting on $\{\delta A, \delta G_o\}$. In short,

we apply the chain rule of differentiation to each of the steps

i)-ix) defining K_o.

δi) given $\{\delta A, \delta G_o\}$, the variation of Γ, called $\delta\Gamma$, is given

by

$$\delta\Gamma(z) = \delta G_o(G_o(z)) + G_o'(G_o(z))\delta G_o(z) , \qquad (5.1.10)$$

δii) solve the equation (obtained by linearizing

$(F_o + \delta F_o) \circ (\Gamma + \delta\Gamma) - (F_o + \delta F_o) = -2)$

$$\delta F_o(\Gamma(z)) - \delta F_o(z) = -F_o'(\Gamma(z))\delta\Gamma(z) , \qquad (5.1.11)$$

δiii) define

$$\delta h(z) = \delta F_o(T_{-A}(-x_{oo} + z)) +$$

$$(5.1.12)$$

$$+ F_o'(T_{-A}(-x_{oo} + z)) \cdot (\delta_A [T_{-A}(-x_{oo} + z)]) \cdot \delta A ,$$

$-$ (same expression at $z = 0$)

δiv) the variation of τ, called $\delta\tau$, is now

$$\delta\tau = -\tau \cdot \log\tau \cdot \delta h'(\tau x_{oo})/\{(\log\tau + 1)h'(\tau x_{oo})$$

$$+ \tau\log\tau \cdot x_{oo}h''(\tau x_{oo})\} , \qquad (5.1.13)$$

δv) now

$$\delta\alpha = \alpha \cdot \{\frac{\delta\tau}{\tau} \cdot h(\tau x_{oo}) + \log\tau \cdot h'(\tau x_{oo}) \cdot x_{oo} \delta\tau$$

$$+ \log\tau \cdot \delta h(\tau x_{oo})\} , \tag{5.1.14}$$

δvi) finally

$$\delta\widetilde{G}(z) = (\widetilde{G}(z) + x_{oo}) \cdot \{\frac{\delta\alpha}{\alpha} - (\frac{1}{\tau} \cdot h(\tau x_{oo} + \tau z)$$

$$+ \log\tau \cdot h'(\tau x_{oo} + \tau z) \cdot (x_{oo} + z))\delta\tau$$

$$- \log\tau \cdot \delta h(\tau x_{oo} + \tau z)\} , \tag{5.1.15}$$

δvii) now

$$\delta\widetilde{A} = \frac{x_{oo}}{4}\{(\tau \cdot h''(\tau x_{oo}) + 2\tau \cdot \log\tau \cdot h''(\tau x_{oo})$$

$$+ \log\tau \cdot \tau^2 \cdot x_{oo} \cdot h'''(\tau x_{oo}))\delta\tau$$

$$+ \log\tau \cdot \tau^2 \cdot \delta h''(\tau x_{oo})\} , \tag{5.1.16}$$

δviii) define

$$\delta\widetilde{G}_o(z) = \{\delta_{\widetilde{A}}[T_{-\widetilde{A}}] \circ \widetilde{G} \circ T_{\widetilde{A}}(z)$$

$$+ T_{-\widetilde{A}}' \circ \widetilde{G} \circ T_{\widetilde{A}}(z) \cdot \widetilde{G}' \circ T_{\widetilde{A}}(z) \cdot \delta_{\widetilde{A}}[T_{\widetilde{A}}(z)]\} \cdot \delta\widetilde{A}$$

$$+ T_{-\widetilde{A}}' \circ \widetilde{G} \circ T_{\widetilde{A}}(z) \cdot \delta\widetilde{G} \circ T_{\widetilde{A}}(z) , \tag{5.1.17}$$

δix) we now have

$$DK_{\{A,G_o\}}(\{\delta A, \delta G_o\}) = \{\delta A, \delta G_o\} . \qquad (5.1.18)$$

Note that except for step δii) all operations are trivial mathematics.

In the next section we shall define a function space on which K_o has a nonempty domain of definition. We then discuss the operator K, the realization of K_o on this space.

5.2. Definition of function spaces

We now fill in the necessary details to make the fixed point problem mathematically well defined and implementable on a computer. (See Section 4 and [16] for more information concerning the method.)

All functions which are handled by the computer program are supposed to be analytic on the unit disk $D = \{z \in \mathbb{C} \mid |z| < 1\}$. All functions which we want to control on the computer have therefore to be analytic on certain disks and are then brought to normal form by a linear change of coordinates. We use the superscript u to indicate that this change of coordinates has been performed.

For our purposes, we fix, guided by trial and error, parameters describing domains as follows :

$$G_o(z) = \rho_o G_o^u(\frac{z}{\rho_o}) , \qquad \rho_o = 0.726 , \qquad (5.2.1)$$

$$\Gamma(z) = \rho_\Gamma \Gamma^u(\frac{z}{\rho_\Gamma}) , \qquad \rho_\Gamma = 0.53 , \qquad\qquad (5.2.2)$$

$$h(z) = h^u(\frac{z - \alpha_h}{\rho_h}) , \quad \alpha_h = 0.024764 , \quad \rho_h = 0.035 , \qquad (5.2.3)$$

$$\tilde{G}(z) = \rho_G \cdot \tilde{G}^u(\frac{z - \alpha_G}{\rho_G}) + \alpha_G , \quad \alpha_G = 0.3507359 , \quad \rho_G = 1.0455. \ (5.2.4)$$

Thus, e.g., h is analytic on a disk of center α_h and radius ρ_h.

We shall also need the sectors

$$S^{\pm} = \{z = \pm\rho e^{i\varphi} | \ |\varphi| < \varphi_o , \ 0 < \rho < \rho_\Gamma\} , \qquad\qquad (5.2.5)$$

and $\varphi_o = 0.135,$

$$(S)^2 = \{z \in \mathbb{C}_2 | \ \frac{4z}{\kappa} \in S^+ \ \text{ or } \ \frac{4z}{\kappa} \in S^-\} , \quad \kappa = \frac{-12}{\Gamma'''(0)} .$$

(See below for more information concerning domains and Lemma 5.4.4 for bounds on κ.)

We write B for the Banach space of functions bounded and analytic on $D_o = \{z \in \mathbb{C} | \ |z| < \rho_o\}$, real on real points, equipped with the supremum norm. We also define B_o as the subspace of B consisting of those functions having vanishing first, second and fourth order derivatives at 0, and we denote by B_1 the functions of the form $-z + f, \ f \in B_o$.

We shall work on the computer not on B_1 but in a subspace equipped with a stronger norm. The idea is that we want to write G_o as

$$G_o(z) = -z + \frac{z^3}{\rho_o^2} \cdot Q\left(\frac{z}{\rho_o}\right) \ ,$$

and to use the ℓ^1-norm for the Taylor coefficients of Q at 0.

Formally, given an element $\underline{a} = (a_1, a_2, \ldots)$ of ℓ^1, we associate

with it an element $\{A, G_o\}$ of $\mathbb{R} \oplus B_1$ by

$$A = 2a_1 - 4a_2 \tag{5.2.6}$$

$$G_o(z) = -z + a_2 \frac{z^3}{\rho_o^2} + \sum_{n=3}^{\infty} a_n \frac{z^{n+2}}{\rho_o^{n+1}} \ . \tag{5.2.7}$$

We denote the set of $\{A, G_o\}$'s obtained that way by A_1, and we

equip A_1 with the norm $\sum_{n=1}^{\infty} |a_n|$. Note that A_1 contains any

element $\{A, G_o\}$ of $\mathbb{R} \oplus B_1$ for which G_o is analytic on the

closure of D_o.

For the remainder of this paper the norm of an element of A_1

will always mean the norm of ℓ^1 type just introduced, and A_1

will eventually be identified with ℓ^1.

Later on, we also need to equip the space $\mathbb{R} \oplus B_o$ with this

norm, i.e. if we are formally given an element $\delta\underline{a} = (\delta a_1, \delta a_2, \ldots)$

of ℓ^1 we associate with it an element $\{\delta A, \delta G_o\}$ in the space

$\mathbb{R} \oplus B_o$ by

$$\delta A = 2 \cdot \delta a_1 - 4\delta a_2 \tag{5.2.8}$$

$$\delta G_o = \delta a_2 \frac{z^3}{\rho_o^2} + \sum_{n=3}^{\infty} \delta a_n \frac{z^{n+2}}{\rho_o^{n+1}} \ . \tag{5.2.9}$$

We denote the set of $\{\delta A, \delta G_o\}$ obtained that way by A_o and we equip A_o with the norm $\sum\limits_{n\geq 1} |a_n|$. We shall also eventually identify A_o with ℓ^1.

At this point we complete our discussion of how we represent functions on the computer. We want to describe in the following the action of the operator K_o on A_1. We therefore need on the computer an additional data structure to describe the ℓ^1-sequence associated to an element $\{A, G_o\} \in A_1$ and $\{\delta A, \delta G_o\} \in A_o$. We do this in total analogy to the case of "vectors". For $\underline{a} \in \ell^1$, $\underline{a} = \{a_1, \ldots\}$ we have

i) $a_n \in \mathbb{R}$, $n = 1, 2, \ldots$,

ii) $\sum\limits_{n=1}^{\infty} |a_n| < \infty$,

and we denote $\|\cdot\|$ the above sum. We define Standard Balls in ℓ^1 as follows. We choose $K > 0$ and we then define a standard ball of degree K by giving K scalars (intervals) $a_i \in S$, $i = 1, \ldots, K$ and two non negative reps (numbers) $a^{(G)} \in R$ and $a^{(H)} \in R$. We denote this sequence $\underline{v\emptyset} = \{v\emptyset_1, \ldots, v\emptyset_K, v\emptyset^{(G)}, v\emptyset^{(H)}\}$, and the associated ball in ℓ^1 is defined by

$$\mathcal{L}(\underline{v\emptyset}) = \{\underline{a} \in \ell_1 | \underline{a} = \{a_1, \ldots, a_K\} + \underline{a}_G + \underline{a}_H ,$$

$$a_i \in v\emptyset_i , \quad i = 1, \ldots, K ,$$

$$\|\underline{a}_G\| \leq v\emptyset^{(G)} , \quad \|\underline{a}_H\| \leq v\emptyset^{(H)}$$

$$\underline{a}_G, \underline{a}_H \epsilon \ell_1 \quad, \quad (\underline{a}_H)_i = 0 \quad, \quad i = 1, \ldots, K \} \ .$$

As for vectors we call $\underline{a}_G$ the "general" and $\underline{a}_H$ the "higher order part" of $\underline{a}$.

$\mathcal{L}(\underline{v}\emptyset)$ has a natural decomposition into a finite sequence, a general and a higher order part. If $\underline{v}\emptyset$ is a sequence we denote it in the program $v\ldots\emptyset$.

A sequence is represented as a FORTRAN-array. The correspondence of a sequence $\underline{v}\emptyset$ with an array, say $vK\emptyset$, is

$$vK\emptyset(i) = v\emptyset_i \ , \quad i = 1, \ldots, K \ ,$$

and

$$vK\emptyset(NERR) = [-_c v\emptyset^{(G)}, v\emptyset^{(H)}] \ .$$

(See Section 4.4.2 for comparison with vectors.)

We next list the various operations on sequences.

If we initialize our subroutine package for vectors to the degree $K - 1$, then the subroutines FSETZERO, FADD and FMINUS operate correctly on sequences of degree K. Furthermore, the function rL1NORM also acts correctly (i.e. calculates $\|\cdot\|$), but we use the name rNORM instead (entry rNORM in subroutine rL1NORM).

The main operations on sequences are the subroutine FvKI which controls the action of the operator K and the subroutine

FvDKI, which bounds the linearized operator DK.

We also need a subroutine which associates, using (5.2.6), (5.2.7) a number A and a function G_O to the ℓ^1-sequence corresponding to $\{A,G_O\}\in A_1$. This subroutine is called UNPACK. Its inverse, which associates to the number A and the function G_O a ℓ^1-sequence, using the inverse of (5.2.6), (5.2.7) is called PACK. Both of these subroutines are listed below and are self-explaining in view of the definitions of "vectors", "sequences" and the relations (5.2.6), (5.2.7). The only precaution which has to be taken is that one has to change the "polynomial degree" from $K = N - 2$ to N in UNPACK and conversely from N to $K = N - 2$ in PACK. This is however also clear by virtue of (5.2.6), (5.2.7).

```
      SUBROUTINE UNPACK(vK0,sAAA,yG0)
      INCLUDE BIGN.STUFF
      DIMENSION vK0(1),vG0(1)
      s000=sHALF
      s111=sTWO
      vG0(1)=sZERO
      vG0(2)=sNEG(sONE)
      vG0(3)=sZERO
      DO 1 I=2,N1
      vG0(I+2)=sNEG(vK0(I))
    1 CONTINUE
      rTG=rG(vK0)
C..the error in sAAA is bounded by rTG*max(1/s000,s111/s000)
      sAAA=sSUM(sDIFF(sQUOT(vK0(1),s000)
     *                ,sPROD(vK0(2),sQUOT(s111,s000)))
     *          ,sPROD(sFOUR,DCMPLX(-rTG,rTG)))
      CALL DEG(N+2)
      CALL FCLEAN(vG0)
      vG0(NERR)=vK0(NERR)
      RETURN
      END
```

```
      SUBROUTINE PACK(sAAA,vGO,vKO)
      INCLUDE GIGN.STUFF
      DIMENSION vGO(1),vKO(1)
      sOOO=sHALF
      s111=sTWO
      rTG=rG(vGO)
      vKO(1)=sSUM(sDIFF(sPROD(sAAA,sOOO)
     *                      ,sPROD(vGO(4),s111))
     *             ,sPROD(s111,DCMPLX(-rTG,rTG)))
      DO 1 I=2,N-1
      vKO(I)=sNEG(vGO(I+2))
    1 CONTINUE
      CALL DEG(N-2)
      CALL FCLEAN(vKO)
      vKO(NERR)=vGO(NERR)
      RETURN
      END
```

This discussion completes this introductory section and we give next the equations defining the operators K and DK respectively.

5.3. Equations defining K and DK

Using the notation of Section 5.2, we define the operator K through the following sequence of steps.

i) given $\{A,G_o\} \in A_1$ construct

$$\Gamma^u(z) = \frac{\rho_o}{\rho_\Gamma} G_o^u (G_o^u (\frac{\rho_\Gamma}{\rho_o} z)) , \qquad (5.3.1)$$

ii) show that the equation

$$F_o(\Gamma(z)) - F_o(z) = -2 \qquad (5.3.2)$$

has a solution on $S^{\pm}$, respectively and construct then

$$h_B^u(z) = F_0(-x_{ohB} + \rho_{hB} \cdot z) \tag{5.3.3}$$

- (same expression at $z = 0$),

and

$$h^u(z) = h_B^u(z) \left(\frac{x_{ohB}}{\rho_{hB}} + \frac{\rho_0}{\rho_{hB}} (G_0^u)^{I_B+1} \left(\frac{1}{\rho_0} T_{-A} (-x_{oh} + \rho_h z) \right) \right)$$

- (same expression at $z = 0$), $\tag{5.3.4}$

where $x_{ohB} = .140849214$, $\rho_{hB} = .002$, $I_B = 39$, and
$x_{oh} = x_{oo} - \alpha_h$.

iii) solve the following equation for τ

$$1 - \frac{\tau x_{oo}}{\rho_h} \cdot \log\tau \cdot h^{u'} \left(\frac{\tau x_{oo} - \alpha_h}{\rho_h} \right) = 0, \tag{5.3.5}$$

iv) set $\alpha = x_{oo} \cdot \exp\left(\log\tau \cdot h^u \left(\frac{\tau x_{oo} - \alpha_h}{\rho_h} \right) \right), \tag{5.3.6}$

v) define

$$\tilde{G}^u(z) = \frac{\alpha}{\rho_G} \cdot \exp\left(-\log\tau \cdot h^u \left(\frac{\tau \rho_G}{\rho_h} \cdot z + \frac{\tau(\alpha_G + x_{oo}) - \alpha_h}{\rho_h} \right) \right)$$

$$- \frac{x_{oo} + \alpha_G}{\rho_G}, \tag{5.3.7}$$

vi) determine

$$\tilde{A} = -\frac{1}{4} \cdot \left(\frac{1}{x_{oo}} + x_{oo} \cdot (-\log\tau) \left(\frac{\tau}{\rho_h} \right)^2 h^{u''} \left(\frac{\tau x_{oo} + \alpha_h}{\rho_h} \right) \right), \tag{5.3.8}$$

vii) define

$$\widetilde{G}_o^u = \frac{1}{\rho_o} \cdot T_{-\widetilde{A}}(\alpha_G + \rho_G \widetilde{G}^u(\frac{1}{\rho_G}(T_{\widetilde{A}}(\rho_o z) - \alpha_G))) \ , \tag{5.3.9}$$

viii) define

$$K(\{A,G_o\}) = \{\widetilde{A},\widetilde{G}_o\} \ . \tag{5.3.10}$$

Note that to use the normalization $h^u(0) = 0$ amounts to a different choice of β in (1.8).

A remark concerning the definitions in (5.3.2)-(5.3.4) seems in place. The reader might ask why we first calculate the function h_B^u in (5.3.3) and only afterwards h^u through (5.3.4). Furthermore it might be unclear why h^u is given by (5.3.4) at all. To explain this, we refer to Fig. 5.3.1 and recall in the following the results established in Section 2 :

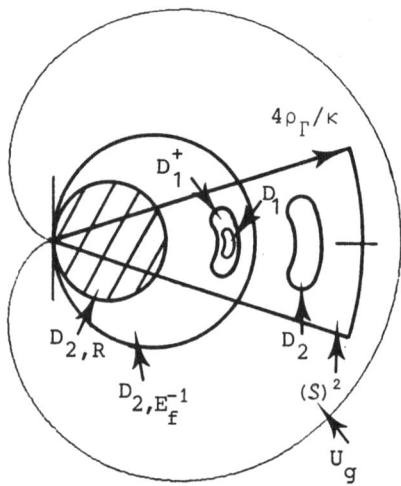

Fig. 5.3.1. Analyticity domains related to (5.3.2)-(5.3.4).

$\mathbb{C}_2$ is the Riemann surface of $z^{1/2}$ and the figure shows the second sheet of this surface. U_g is the domain of points attracted by
the function g, where g is the normal form if the function Γ
in (5.3.2). We shall show that U_g contains the sector $(S)^2$ (see
(5.2.5) for its definition), by proving that $g((S)^2) \subset (S)^2$ and

that $\lim_{\nu \to \infty} g^\nu((S)^2) = 0$. Furthermore, we have shown in Section 2

that U_g contains some disk $D_{2,R}$. It follows that (the normal
form of) (5.3.2) has a solution f_o which is analytic at least on
the union $(S)^2 \cup D_{2,R}$. We have also seen that the asymptotic expansion of f_o at $z = 0$ is Borel summable and defines some function $\tilde{f}_o$ which is analytic on some disk $D_{2,E_f^{-1}}$. (We disregard
in this discussion the fact that one first has to split off the
singular part of f_o.) Furthermore if the radius R of $D_{2,R}$ is
sufficiently small, we have not only $D_{2,R} \subset U_g$ but even

$g(D_{2,R}) \subset D_{2,R} \subset D_{2,E_f^{-1}}$. We conclude that the Borel sum $\tilde{f}_o$ solves
(the normal form of) (5.3.2) on $D_{2,R}$. But we also showed that the
solution of (the normal form of) (5.3.2) is unique, on $D_{2,R}$ say,
and therefore $f_o = \tilde{f}_o$ on this domain. It follows that f_o can
be analytically continued from U_g to $U_g \cup D_{2,E_f^{-1}}$ (which may be
a trivial statement if $D_{2,E_f^{-1}} \subset U_g$) and, more importantly, that
the representation of f_o by its Borel sum solves (the normal form
of) (5.3.2) not only on $D_{2,R}$ but on $D_{2,E_f^{-1}} \cap U_g$ and therefore in
particular on $D_{2,E_f^{-1}} \cap (S)^2$. We shall use this representation to
construct the function f_o on the domain

$D_1 \equiv \{ z \in \mathbb{C}_2 | \sqrt[4]{z} \in D(\alpha_{h_B}, \rho_{h_B}) \}$. Unfortunately the domain

$D_2 \equiv \{ z \in \mathbb{C}_2 | \sqrt[4]{z} \in T_{-A}(D(\alpha_h, \rho_g)) \}$ on which we actually need to con-
trol the function f_o is not contained in the intersection

$D_{2,E_f^{-1}} \cap (S)^2$ but since we know that f_o exists on $(S)^2$ and that

$(S)^2$ is invariant under g and attracted by $z = 0$ we can use

(the normal form of) the functional equation (5.3.2) to express

f_o on D_2 in terms of f_o on D_1. This is exactly what we do

in (5.3.4) (directly for the functions F_o and Γ instead of f_o

and g). The number $I_B = 39$ in (5.3.4) has been chosen large

enough such that $g^{(I_B+1)/2}(D_2) \subset D_1 \subset D_1^+ \subset D_{2,E_f^{-1}}$. The domain

$D_1^+ \supset D_1$ is only used to establish certain ℓ^1-norms on D_1 when

we are given sup-norms on D_1^+. See the discussion of FSUMBOREL.

Remember that we want to use the contraction mapping principle

to prove the existence (and local uniqueness) of a fixed point of

K. Thus, as we have already mentioned in the discussion of the

operators K_{oo} and K_o we need to find a choice of $\{A,G_o\} \in A_1$

for which

$$\| K(\{A,G_o\}) - \{A,G_o\} \| = \varepsilon$$

with ε small (the norm $\| \cdot \|$ is now well defined). Our choice

for $\{A,G_o\}$ is a finite ℓ^1-sequence of rational numbers a_n given

in Table I, and corresponding to a polynomial G_o of degree 67.

These numbers have been obtained by using a very unsophisticated

(numerical) computer program which iterates the operator K_o a

certain number of times (taking as initial choice

$G(z) \approx \frac{1}{e} \cdot (\exp(-e \cdot z) - 1))$, and normalizing the result obtained

corresponding to (1.8) and (1.10). The resulting Taylor coeffi-

cients of G are listed in Table II and correspond to about 30

iterations of the operator K_o. Our choice for x_{oo} corresponds

to a scale on which numerically $\alpha = 1$ and $\beta = 0$ in (1.9) and

(1.8), respectively.

We give ourselves a ball U_β of radius β around $\{A, \mathring{G}_o\}$,

$$U_\beta = \{\{A, G_o\} \epsilon A_1 \,|\, \|\{A, G_o\} - \{\mathring{A}, \mathring{G}_o\}\| < \beta\} \ .$$

We shall show that the operator K is defined as a mapping from U_β into A_1. Furthermore, it will be seen that K is once continuously differentiable as a mapping from U_β to A_1 and that the derivative DK is a compact operator on A_o. In fact, by using the analyticity properties it can be shown that K is infinitely often differentiable.

We next describe the action of the linear operator $DK_{\{A, G_o\}}$. It is given by the following sequence of steps (which are obtained by applying the chain rule of differentiation of (5.3.1)-(5.3.10)) :

$\delta i)$ given $\{\delta A, \delta G_o\} \epsilon A_o$ define $\delta\Gamma^u$ by

$$\delta\Gamma^u(z) = \frac{\rho_o}{\rho_\Gamma} G_o^{u'}(G_o^u(\frac{\rho_\Gamma}{\rho_o}z)) \delta G_o^u(z)$$

$$+ \frac{\rho_o}{\rho_\Gamma} \delta G_o^u(G_o^u(\frac{\rho_\Gamma}{\rho_o}z)) \tag{5.3.11}$$

$\delta ii)$ show that the equation

$$\delta F_o(\Gamma(z)) - \delta F_o(z) = -F_o'(\Gamma(z)) \delta\Gamma(z) \tag{5.3.12}$$

has a solution δF_o on $S^{\pm}$ and construct

$$\delta h_B^u(z) = \delta F_o(-x_{oh_B} + \rho_{h_B} z) - \delta F_o(-x_{oh_B}) , \qquad (5.3.13)$$

and

$$\delta h^u(z) = \delta h_B^u(\frac{x_{oh}}{\rho_{h_B}} + \frac{\rho_o}{\rho_{h_B}} \cdot (G_o^u)^{I_B+1} (\frac{1}{\rho_o} T_{-A}(-x_{oh} + \rho_h \cdot z)))$$

$$+ h_B^{u'}(\frac{x_{oh}}{\rho_{h_B}} + \frac{\rho_o}{\rho_{h_B}} \cdot (G_o^u)^{I_B+1} (\frac{1}{\rho_o} T_{-A}(-x_{oh} + \rho_h \cdot z)))$$

$$\cdot \{\frac{\rho_o}{\rho_{h_B}} \cdot \sum_{i=1}^{I_P+1} [\delta G_o^u((G_o^u)^{I_B+1-i} (\frac{1}{\rho_o} T_{-A}(-x_{oh} + \rho_h \cdot z)))$$

$$\prod_{j=1}^{i-1} \cdot G_o^{u'}((G_o^u)^{I_B+1-j} (\frac{1}{\rho_o} T_{-A}(-x_{oh} + \rho_h \cdot z)))]$$

$$+ \frac{1}{\rho_{h_B}} \cdot \prod_{j=1}^{I_B+1} G_o^{u'}((G_o^u)^{I_B+1-j}) (\frac{1}{\rho_o} T_{-A}(-x_{oh} + \rho_h \cdot z))$$

$$\cdot \delta_A T_{-A}(-x_{oh} + \rho_h z) \cdot \delta A\} \qquad (5.3.14)$$

- (same expression at $z = 0$)

δiii) the variation of τ, called $\delta\tau$, is now

$$\delta\tau = -\frac{\tau x_{oo}}{\rho_h} \cdot \log\tau \cdot \delta h^{u'}(\frac{\tau x_{oo}}{\rho_h}^{-\alpha}h) \cdot \qquad (5.3.15)$$

$$\cdot [\frac{x_{oo}}{\rho_h} \cdot (\log\tau+1) h^{u'}(\frac{\tau x_{oo}}{\rho_h}^{-\alpha}h) + \frac{\tau x_{oo}^2}{\rho_h^2} \cdot \log\tau \cdot h^{u''}(\frac{\tau x_{oo}}{\rho_h}^{-\alpha}h)]^{-1} ,$$

δiv) and the variation of α is

$$\delta\alpha = \alpha[\frac{\delta\tau}{\tau} \cdot h(\frac{\tau x_{oo}}{\rho_h}^{-\alpha}h) + \delta\tau \cdot \log\tau \cdot h^{u'}(\frac{\tau x_{oo}}{\rho_h}^{-\alpha}h) \frac{x_{oo}}{\rho_h}$$

$$+ \log\tau \cdot \delta h^u(\frac{\tau x_{oo}^- h}{\rho_h})] . \qquad (5.3.16)$$

δv) this leads to

$$\delta\widetilde{G}^u(z) = \frac{\delta\alpha}{\alpha}\widetilde{G}^u(z) + \widetilde{G}^{u'}(z)\cdot[-\frac{\delta\tau}{\tau}\cdot h^u(\frac{\tau\rho_G}{\rho_h}\cdot z + \frac{\tau(\alpha_G+x_{oo})-\alpha_h}{\rho_h})$$

$$- \log\tau\cdot h^{u'}(\frac{\tau\rho_G}{\rho_h}\cdot z + \frac{\tau(\alpha_G+x_{oo})-\alpha_h}{\rho_h})\cdot(\frac{\delta\tau\cdot(\rho_G\cdot z+\alpha_G+x_{oo})}{\rho_h})$$

$$- \log\tau\cdot\delta h^u(\frac{\tau\rho_G}{\rho_h}\cdot z + \frac{\tau\cdot(\alpha_G+x_{oo})}{\rho_h})] \quad, \tag{5.3.17}$$

δvi) now

$$\delta\widetilde{A} = \frac{1}{4}x_{oo}[(\frac{\delta\tau}{\tau}(\frac{\tau}{\rho_h})^2 + \log\tau\cdot\frac{2\tau}{\rho_h^2}\cdot h^u(\frac{\tau x_{oo}-\alpha_h}{\rho_h}) + \log\tau\cdot(\frac{\tau}{\rho_h})^2$$

$$\cdot(\delta h^u(\frac{\tau x_{oo}-\alpha_h}{\rho_h}) + h^u(\frac{\tau x_{oo}-\alpha_h}{\rho_h})\frac{\delta\tau\cdot x_{oo}}{\rho_h})] \quad, \tag{5.3.18}$$

δvii) and finally

$$\delta\widetilde{G}^u_o(z) = -[\frac{\alpha_G}{\rho_o} + \frac{\rho_G}{\rho_o}\cdot\widetilde{G}^u(\frac{1}{\rho_G}T_A(\rho_o z) - \frac{\alpha_G}{\rho_G})]$$

$$\cdot[1 - A\alpha_G - A\rho_G\widetilde{G}^u(\frac{1}{\rho_G}T_A(\rho_o z) - \frac{\alpha_G}{\rho_G})]^{-2}$$

$$\cdot[-A\rho_G\delta\widetilde{G}^u(\frac{1}{\rho_G}T_A(\rho_o z) - \frac{\alpha_G}{\rho_G})$$

$$+ (-\alpha_G - \rho_G\widetilde{G}^u(\frac{1}{\rho_G}T_A(\rho_o z) - \frac{\alpha_G}{\rho_G})$$

$$+ A\rho_o\widetilde{G}^{u'}(\frac{1}{\rho_G}T_A(\rho_o z) - \frac{\alpha_G}{\rho_G})\frac{\rho_o z^2}{(1+A\rho_o z)^2})\delta A] + \tag{5.3.19}$$

$$+ \; [\frac{\rho_G}{\rho_o}\delta\widetilde{G}^u(\frac{1}{\rho_G}T_A(\rho_o z) \; - \; \frac{\alpha_G}{\rho_G})$$

$$- \; \widetilde{G}^{u'}(\frac{1}{\rho_G}T_A(\rho_o z) \; - \; \frac{\alpha_G}{\rho_G})\frac{\rho_o z^2}{(1+A\rho_o z)^2}\delta A]$$

$$\cdot \; [1 \; - \; A\alpha_G \; - \; A\rho_G\widetilde{G}^u(\frac{1}{\rho_G}T_A(\rho_o z) \; - \; \frac{\alpha_G}{\rho_G})] \;\; ,$$

δviii) and therefore we get

$$DK_{\{A,G_o\}}(\{\delta A,\delta G_o\}) \; = \; \{\delta\widetilde{A},\delta\widetilde{G}_o\} \; . \tag{5.3.20}$$

Note that except for (5.3.13) all steps are trivial mathematics. The advantage of propagating the estimates in an automatic fashion (namely assisted by a computer) is obvious.

This completes our discussion of the equations defining the operators K and DK. In the next section we present our results.

5.4. Statement of results

In this section we state our results. The estimates necessary for the proof of these results are obtained by running our computer program which we discuss in the subsequent sections.

Theorem 5.4.1. There is a ball $D_K \subset A_1$ on which K is defined.

Theorem 5.4.2. The operator K has a fixed point $\{A^*,G_o^*\}$ in $D_K \subset A_1$. It is unique in D_K.

The necessary bounds for these theorems are obtained by running the computer program. We give $\{A, G_o\}$ as described above and choose as the ball D_K the ball $U_\beta\big|_{\beta=6.10^{-10}}$, where

$$U_\beta = \{\{A, G_o\} \in A_1 \mid \|\{A, G_o\} - \{\hat{A}, \hat{G}_o\}\| < \beta\} .$$

Theorem 5.4.1 follows from the fact that there is no "domain error" occurring when we are running the program with the choice $\beta = 6.10^{-10}$ i.e. all operators lead to functions and evaluations which are legal within the (somewhat too restrictive) set of rules we have described in Section 4. (We choose for the polynomial degree $N = 50$ for this run. See below for details.) Theorem 5.4.2 follows from the contraction mapping principle, using the fact that the numbers ε, β, ρ in the following result satisfy $0 < \varepsilon < (1-\rho)\beta$. ($\varepsilon$ has been obtained for a polynomial degree $N = 67$ and ρ with degrees 65 and 50 respectively. See below for details.)

Theorem 5.4.3. <u>The operator K is defined on $U_\beta\big|_{\beta=6.10^{-10}}$ and maps it to A_1. We have the following estimates</u>

1) <u>one has the bound</u>

$$\|K(\{A, G_o\}) - \{A, G_o\}\| < \varepsilon = 0.1220127 \cdot 10^{-9}$$

2) <u>the tangent map DK has operator norm bounded by</u>

$$\|DK_{\{A, G_o\}}\| < \rho = 0.782599$$

for every $\{A,G_o\} \in U_\beta \big|_{\beta=6.10^{-10}}$

For completeness, we also give the following bound.

<u>Lemma 5.4.4.</u> <u>Let</u>$\{A,G_o\} \in D_K$. <u>By definition, the function</u> G_o <u>is of the form</u> $G_o(z) = -z(1-z^2/\kappa + O(z^3))$. <u>We have the</u> <u>bound</u> $1.813 < \kappa < 1.814$.

5.5. The main program

In this section we start disucssing Theorem 5.4.3, i.e. we begin a detailed study of our computer program. We essentially follow the equations (5.3.1)-(5.3.10) and (5.3.11)-(5.3.20) which define the operators K and DK and discuss in detail the bounds which need to be established.

We first give the listing of the main program, and then the include-files STUFF and STRES1. (The meaning of an include file is that it gets substituted at compile-time for the corresponding include-statements in a program.)

The main program

```
 1:        INCLUDE BIGN.STUFF
 2:        COMMON/DOMAIN/sRHOGO,sRHOGAMMA,sXOh,sCENTh,sRHOh,sCENTG,sRHOG
 3:       *           ,sRHOhB,sRHOPhB,sXOhB,sPHIO,sxKAPPA,sKAPPA
 4:       *           ,sRHOP,sRHO1
 5:        COMMON/SCALE /sXOO
 6:        COMMON/LOFFEL/sBETA,xPHI(NDEG),vINT1(NHALF,NDEG)
 7:       *           ,vINT2(NHALF,NDEG)
 8:        COMMON/ECALLE/sC3
 9:        COMMON/REDUCE/sGO3,sGO4,sGO5,sGA3,sGA33,sGA5,sGA6
10:        COMMON/SWITCH/ITEST
11:        INCLUDE BIGN.STRES1
12:        DIMENSION vKO(NDEG),vKOI(NDEG)
13:        DIMENSION vDKO(NDEG),vDKOI(NDEG)
14:        DIMENSION vTEMPO(NDEG)
15:        DIMENSION vDZh(NDEG)
16:        DIMENSION vT1(NDEG),vT2(NDEG),vT3(NDEG)
17:        DIMENSION xT1(NDEG)
18:
19:C..activate exponent overflow test
20:        CALL OVFSET(10000)
21:
22:
23:C   Initialize parameters for subroutines to degree N
24:C   =================================================
25:
26:        READ(5,*)N
27:        WRITE(6,*)'N=        ',N
28:        CALL INITIALIZE(N-2)
29:
30:
31:C   Set scale
32:C   =========
33:
34:        sXOO=saCONST(.3911329993510225420+00)
35:
36:
37:C   Choose domains
38:C   ==============
39:
40:C   Choose domains in z-plane
41:C   -------------------------
42:C..the following scalars contain only one rep
43:        sRHOGO    =saCONST(.7260000000+00)
44:        sRHOGAMMA=saCONST(.5300000000+00)
45:        sXOhB     =saCONST(.1408492140+00)
46:        sRHOhB    =saCONST(.0020000000+00)
47:        sRHOPhB   =saCONST(.0080000000+00)
48:        sPHIO     =saCONST(.1350000000+00)
49:        sCENTh    =saCONST(.0247640000+00)
50:        sRHOh     =saCONST(.0350000000+00)
51:        sCENTG    =saCONST(.3507359000+00)
52:        sRHOG     =saCONST(1.0455000000+00)
53:C..the following scalars are only modulo up and down localized
54:        sXOh      =sDIFF(sXOO,sCENTh)
55:        sRHOP     =sSUM(sXOhB,sRHOPhB)
56:        sRHO1     =sQUOT(sRHOP,sRHOGAMMA)
57:
58:C   Choose domain in t-plane; set conformal mapping
59:C   -----------------------------------------------
60:C..bound for t**m/((1-exp(t))*m!)
61:        sC3=saCONST(1.8484000+00)
62:C..choose radius for Borel summation using Loeffel's method
63:        sBETA=saCONST(.9065000+00)
64:C..set conformal mapping (calculate coefficients up to the maximal degree)
65:        NSAVE=N
```

```
66:          CALL DEG(NHALF+5)
67:          CALL CSETPHI(xPHI)
68:          CALL DEG(NSAVE)
69:
70:
71:C   Read input vector
72:C   =================
73:
74:C..set vector equal to zero
75:          CALL FSETZERO(vK0)
76:C..read vector
77:          DO 1 I=1,N1
78:          READ(18,77)aT1
79:          vK0(I)=saCONST(aT1)
80:    1 CONTINUE
81:   77 FORMAT(D32.24)
82:
83:
84:C   Remark: We define the binary content of vK0 to be the
85:C   ------  input vector
86:
87:
88:
89:
90:
91:C   ============================
92:C   Calculate l-norm of vK0-VK0I
93:C   ============================
94:
95:
96:
97:
98:C   Preparatory calculation
99:C   =======================
100:
101:C   Calculate power series xGAMMA
102:C   -----------------------------
103:          NSAVE=N
104:          CALL UNPACK(vK0,sT1,xT1)
105:          CALL CCLEAN(xT1)
106:          CALL CSHFTL(xT1,xT1)
107:          ITEST=0
108:          CALL FsDILAT(xT1,sINV(sRHOG0),xT1)
109:          ITEST=1
110:          CALL CSHFTR(xT1,xT1)
111:C..calculate sxKAPPA
112:          sxKAPPA=sINV(xT1(4))
113:C..continue calculation of xGAMMA
114:          CALL CCOMP(xT1,xT1,xGAMMA)
115:          xGAMMA(1)=sZERO
116:          xGAMMA(2)=sONE
117:          xGAMMA(3)=sZERO
118:          xGAMMA(4)=sNEG(sPROD(sTWO,xT1(4)))
119:          xGAMMA(5)=sZERO
120:          CALL DEG(NSAVE)
121:
122:
123:C   Calculate image vector
124:C   ======================
125:
126:          CALL FvKI(vK0,vK0I,0)
127:
128:
129:C   Calculate norm of difference. Produce output for information
130:C   ============================================================
```

```
131:
132:        CALL FEQUAL(vK0I,vTEMP0)
133:        CALL FMINUS(vK0,vTEMP0)
134:        rEPSILON=rNORM(vTEMP0)
135:        WRITE(6,*)´rEPSILON=´,rEPSILON
136:
137:
138:
139:
140:
141:
142:
143:
144:C   =====================
145:C   Compute tangent map DK
146:C   =====================
147:
148:
149:
150:
151:C   Fix radius of ball in function space. Produce output for information
152:C   ====================================================================
153:
154:        rBETA=6.D-10
155:        WRITE(6,*)´rBETA=´,rBETA
156:
157:
158:C   Make ball of input vectors
159:C   ==========================
160:
161:        vK0(NERR)=error(rBETA,rZERO)
162:
163:
164:C   Prepare loop over basis vectors: put sup over norms equal to zero
165:C   ================================================================
166:
167:        rRHO=rZERO
168:
169:
170:C   We compute tangent map in lower degree for higher indices
171:C   =========================================================
172:
173:        DO 11 JPART=1,2
174:
175:
176:C   Reduce degree
177:C   =============
178:
179:        N1SAVE=N1
180:        READ(5,*)NT
181:        WRITE(6,*)´degree for tangent map=´,NT
182:        CALL DEG(NT-2)
183:        sTH=sCONST(rH(vK0))
184:        DO 2 I=N2,N1SAVE
185:        sTH=sSUM(sTH,sABS(vK0(I)))
186:        vK0(I)=sZERO
187:      2 CONTINUE
188:        vK0(NERR)=error(rG(vK0),rMAXABS(sTH))
189:
190:
191:C   Calculate ball-dependent parts of COMMON/RES1  /; part 1
192:C   ========================================================
193:
194:        CALL FvKI(vK0,vK0I,1)
195:
```

```
196:
197:C  Calculate ball-dependent parts of COMMON/RES1  /; part 2
198:C  ==========================================================
199:
200:C  Initialize to degree of functions
201:C  ---------------------------------
202:      NSAVE=N
203:      CALL DEG(N+2)
204:
205:C  Calculate function vDZG0G0
206:C  --------------------------
207:      CALL FDZCOMP(vG0,vG0RED,vDZG0G0)
208:      vDZG0G0(1)=sNEG(sONE)
209:
210:C  Calculate bounds sCQ and sEQ
211:C  ----------------------------
212:      sRHO=saCONST(.450000D+00)
213:C..calculate sEQ
214:      sEQ=sUPPER(sQUOT(sxKAPPA,sPROD(sPROD(sRHO,sRHO),sFOUR)))
215:C..calculate sCQ
216:      CALL FEQUAL(vG0,vT1)
217:      vT1(NERR)=sZERO
218:      CALL FDZsDILAT(vT1,sQUOT(sRHO,sRHOG0),vT2)
219:      CALL FINVERS(vT2,vT2)
220:      CALL FsDILAT(vT1,sQUOT(sRHO,sRHOG0),vT3)
221:      CALL FDZCOMP(vT1,vT3,vT3)
222:      CALL FINVERS(vT3,vT3)
223:      CALL FMULT(vT2,vT3,vT1)
224:      vT1(1)=sZERO
225:      vT1(2)=sZERO
226:      vT1(3)=sZERO
227:      vT1(4)=sZERO
228:      rT1=rL1NORM(vT1)
229:      sCQ=sUPPER(sPROD(sPROD(sEQ,sEQ),sCONST(rT1)))
230:
231:C  Compute bounds sCJ and sEJ
232:C  --------------------------
233:C..compute sEJ
234:      sEJ=sEF0
235:C..compute sCJ
236:      sRHO=sSQRT(sQUOT(sxKAPPA,sPROD(sFOUR,sEJ)))
237:      CALL FsDILAT(vxJ,sQUOT(sRHO,sRHOGAMMA),vT1)
238:      rT1=rL1NORM(vT1)
239:      sCJ=sUPPER(sPROD(sEJ,sCONST(rT1)))
240:
241:
242:C  Calculate ball-dependent parts of COMMON/RES1  /; part 3
243:C  ==========================================================
244:
245:C  (Calculate bounds sGDZDF0 and sHDZDF0)
246:C  ======================================
247:
248:C  Estimate abs(vDZF0(vGAMMA+sGGAMMA+sHGAMMA)-vDZF0(vGAMMA))
249:C  --------------------------------------------------------
250:      sT1=sQUOT(sSUM(sGGAMMA,sHGAMMA),sPROD(sETA,sRHOP))
251:      IF(DIMAG(sDIFF(sONE,sT1)).LT.rZERO)WRITE(6,*)´error: sT1=´,sT1
252:      sT2=sPROD(sPROD(sINV(sPOWER(sETA,2))
253:     *   ,sQUOT(sSUM(sTWO,sT1),sPOWER(sDIFF(sONE,sT1),2))),sT1)
254:      sT3=sDIFF(sINV(sZ),sQUOT(sPOWER(sRHOP,2),sR))
255:      IF(DIMAG(sT3).LT.rZERO)WRITE(6,*)´error: denominator 1 negative´
256:     *                     ,sT3
257:      sT3=sINV(sPOWER(sT3,2))
258:      sT4=sSUM(sSUM(sTHREE,sPROD(sTHREE,sT1)),sPOWER(sT1,2))
259:      sT5=sQUOT(sPROD(sT3,sT4)
260:     *   ,sPROD(sPOWER(sETA,4),sPOWER(sDIFF(sONE,sT1),3)))
```

```
261:        sT6=sDIFF(sINV(sZ),sSUM(sT2,sQUOT(sPOWER(sRHOP,2),sR)))
262:        IF(DIMAG(sT6).LT.rZERO)WRITE(6,*)'error: denominator 2 negative'
263:     *                                  ,sT6
264:        sT7=sQUOT(sTWO,sPOWER(sT6,3))
265:        sT8=sQUOT(sPROD(sSUM(sTWO,sT1),sT7)
266:     *    ,sPROD(sPOWER(sETA,6),sPOWER(sDIFF(sONE,sT1),5)))
267:        sT9=sPROD(sPROD(sTWO,sC),sSUM(sT5,sT8))
268:C..calculate general
269:        sGJDZF=sUPPER(sPROD(sT9,sGGAMMA))
270:C..calculate higher order
271:        sHJDZF=sUPPER(sPROD(sT9,sHGAMMA))
272:
273:C   Isolate sFP00,sFP01,sFP02
274:C   -------------------------
275:        CALL CDERIVATE(xF0,xT1)
276:        sFP00=xT1(2)
277:        sFP01=xT1(3)
278:        sFP02=xT1(4)
279:
280:C   Isolate sB0,sB1,sB2 and calculate sCT
281:C   -------------------------------------
282:        sB0=xF0(3)
283:        sB1=sQUOT(xF0(4),sPROD(sSQRT(sPI),sHALF))
284:        sB2=xF0(5)
285:        sCT=sSUM(sC,sSUM(sABS(sB0),sSUM(sPROD(sABS(sB1)
286:     *    ,sSQRT(sQUOT(sR,sPROD(sTWO,sEXP(sONE))))),sPROD(sABS(sB2)
287:     *    ,sQUOT(sR,sEXP(sONE))))))
288:        sTRO=sPROD(sBETA,sLOG(sTWO))
289:        sCTT=sPROD(sTOs(sINV(sTRO),s3HALF),sPROD(sEXP(sQUOT(sTRO,sR))
290:     *    ,sCT))
291:
292:C   Compute bound sDZF0GAMMA
293:C   ------------------------
294:        sDZF0GAMMA=sPROD(sTWO,sPROD(sC,sPROD(sQUOT(sRHOP,sPOWER(sETA,3))
295:     *          ,sT3)))
296:
297:C   Compute bounds sGDZGAMMA and sHDZGAMMA
298:C   --------------------------------------
299:        rRHO1=rMAXABS(sRHO1)
300:        sGDZGAMMA=sPROD(sCONST(rG(vGAMMA)),sCONST(rSUP(rRHO1, 3,1)))
301:        sHDZGAMMA=sPROD(sCONST(rH(vGAMMA)),sCONST(rSUP(rRHO1,N1,1)))
302:
303:C   Compute bound sGAMMA
304:C   --------------------
305:        CALL FsDILAT(vGAMMA,sRHO1,vT1)
306:        sGAMMA=sPROD(sCONST(rL1BOUND(vT1)),sRHOGAMMA)
307:
308:C   Compute bound sDZGAMMA
309:C   ----------------------
310:        CALL FEQUAL(vGAMMA,vT1)
311:        CALL FDZsDILAT(vT1,sRHO1,vT1)
312:        sDZGAMMA=sCONST(rL1BOUND(vT1))
313:
314:C   Compute bound s2GAMMA
315:C   ---------------------
316:        CALL FsDILAT(vGAMMA,sRHO1,vT1)
317:        CALL FMULT(vT1,vT1,vT1)
318:        s2GAMMA=sPROD(sCONST(rL1BOUND(vT1)),sPOWER(sRHOGAMMA,2))
319:
320:C   Compute bound sGHGAMMA
321:C   ----------------------
322:        sGHGAMMA=sSUM(sGGAMMA,sHGAMMA)
323:
324:C   Estimate abs(vDZ(GF0+HF0)): calculate delta(DZJ).tot
325:C   ----------------------------------------------------
```

```
326:C..calculate sTGJDZF and sTDZFOGAMMA
327:        sT3=sDIFF(sINV(sZ),sQUOT(sPOWER(sRHOP,2),sR))
328:        sT3=sINV(sPOWER(sSQRT(sT3),7))
329:        sT5=sQUOT(sPROD(sT3,sT4)
330:     *     ,sPROD(sPOWER(sETA,4),sPOWER(sDIFF(sONE,sT1),3)))
331:        sT7=sQUOT(sQUOT(siCONST(7),sTWO),sPOWER(sSQRT(sT6),9))
332:        sT8=sQUOT(sPROD(sSUM(sTWO,sT1),sT7)
333:     *     ,sPROD(sPOWER(sETA,6),sPOWER(sDIFF(sONE,sT1),5)))
334:        sFAC=sPROD(sQUOT(siCONST(5),sTWO),sPROD(s3HALF,sPROD(sHALF
335:     *     ,sSQRT(sPI))))
336:        sT9=sPROD(sPROD(sPROD(sTWO,sFAC),sCTT),sSUM(sT5,sT8))
337:        sTGJDZF=sUPPER(sPROD(sPROD(sPOWER(sRHOP,3),sT9),sGGAMMA))
338:        sTDZFOGAMMA=sPROD(sTWO,sPROD(sPROD(sCTT,sFAC)
339:     *              ,sPROD(sQUOT(sPOWER(sRHOP,4),sPOWER(sETA,3)),sT3)))
340:C..calculate coefficients sJ1, sJ2 and sJ3
341:        sT1=sQUOT(sGGAMMA,sPOWER(sRHOP,3))
342:        sT1=DCMPLX(-rMAXABS(sT1),rMAXABS(sT1))
343:        sT2=sQUOT(sT1,sPOWER(sRHOGAMMA,2))
344:        sJ1H=sPROD(sFPOO,sPROD(sFOUR,sT1))
345:        sJ1 =sSUM(sJO1,sJ1H)
346:        sJ2H=sPROD(sFPO1,sPROD(siCONST(5),sT1))
347:        sJ2 =sSUM(sJO2,sJ2H)
348:        sJ3H=sSUM(sPROD(sFPOO,sSUM(sPROD(sT1,sQUOT(siCONST(12)
349:     *     ,sKAPPA)),sPROD(sT2,siCONST(6)))),sPROD(sFPO1
350:     *     ,sPROD(siCONST(6),sT1)))
351:        sJ3 =sSUM(sJO3,sJ3H)
352:C..calculate bound for general
353:        sT1=sSUM(sSUM(sSUM(sGDZJ,sPROD(sSUM(sPROD(sABS(sFPOO),sGAMMA)
354:     *     ,sSUM(sPROD(sABS(sFPO1),s2GAMMA),sPROD(sABS(sFPO2)
355:     *     ,sPROD(sGAMMA,s2GAMMA)))),sGDZGAMMA))
356:     *     ,sSUM(sPROD(sTDZFOGAMMA,sGDZGAMMA)
357:     *     ,sPROD(sSUM(sABS(sFPOO),sSUM(sPROD(sABS(sFPO1),sSUM(sGAMMA
358:     *     ,sGAMMA)),sPROD(sABS(sFPO2),sSUM(sPROD(sTHREE,s2GAMMA)
359:     *     ,sSUM(sPROD(sTHREE,sPROD(sGAMMA,sGHGAMMA))
360:     *     ,sPOWER(sGHGAMMA,2))))),sPROD(sDZGAMMA,sGGAMMA))))
361:     *     ,sPROD(sTGJDZF,sDZGAMMA))
362:        sT2=sSUM(sSUM(sPROD(sJ1H,sPOWER(sRHOP,3))
363:     *     ,sPROD(sJ2H,sPOWER(sRHOP,4))),sPROD(sJ3H
364:     *     ,sPOWER(sRHOP,5)))
365:        sGDZJTOT=sCONST(rMAXABS(sSUM(sT1,sT2)))
366:C..calculate bound for higher order
367:        sT1=sSUM(sSUM(sHDZJ,sSUM(sPROD(sDZFOGAMMA,sHDZGAMMA)
368:     *     ,sPROD(sSUM(sABS(sFPOO),sSUM(sPROD(sABS(sFPO1)
369:     *     ,sSUM(sGAMMA,sGAMMA)),sPROD(sABS(sFPO2),sSUM(sPROD(sTHREE
370:     *     ,s2GAMMA),sSUM(sPROD(sTHREE,sPROD(sGAMMA,sGHGAMMA))
371:     *     ,sPOWER(sGHGAMMA,2)))))),sPROD(sDZGAMMA,sHGAMMA))))
372:     *     ,sPROD(sQUOT(sCT,sC),sPROD(sHJDZF,sDZGAMMA)))
373:        sHDZJTOT=sCONST(rMAXABS(sT1))
374:C..invert Ecalle's equation: lower order coefficients
375:        sDZFO1=sQUOT(sJ1,sPROD(sFOUR,sGA3))
376:        sDZFO2=sQUOT(sJ2,sPROD(siCONST(5),sGA3))
377:        sDZFO3=sQUOT(sDIFF(sJ3,sPROD(sSUM(sPROD(s3HALF
378:     *          ,sQUOT(sGA5,sGA3)),sPROD(sQUOT(sTHREE,sFOUR),sGA3))
379:     *          ,sJ1)),sPROD(siCONST(6),sGA3))
380:C..add additional contribution to general
381:        CALL FsDILAT(vGAMMA,sRHO1,vT1)
382:        CALL FsMULT(vT1,sRHOGAMMA,vT1)
383:        CALL FDZsDILAT(vGAMMA,sRHO1,vT2)
384:        CALL FMULT(vT1,vT2,vT2)
385:        CALL FEQUAL(vT2,vT3)
386:        vT3(2)=sZERO
387:        sADD1=sCONST(rL1NORM(vT3))
388:        CALL FMULT(vT1,vT2,vT2)
389:        CALL FEQUAL(vT2,vT3)
390:        vT3(3)=sZERO
```

```
391:        sADD2=sCONST(rL1NORM(vT3))
392:        CALL FMULT(vT1,vT2,vT3)
393:        vT3(4)=sZERO
394:        sADD3=sCONST(rL1NORM(vT3))
395:        sT1=sSUM(sPROD(sABS(sDZF01),sADD1),sSUM(sPROD(sABS(sDZF02)
396:     *     ,sADD2),sSUM(sPROD(sABS(sDZF03),sADD3),sSUM(sSUM(sPROD(sABS(sJ1),sPOWER(sRHOP,3))
397:     *     ,sSUM(sSUM(sPROD(sABS(sJ1),sPOWER(sRHOP,3))
398:     *     ,sPROD(sABS(sJ2),sPOWER(sRHOP,4))),sPROD(sABS(sJ3)
399:     *     ,sPOWER(sRHOP,5))))))
400:        sGDZJTOT=sSUM(sGDZJTOT,sCONST(rMAXABS(sT1)))
401:
402:C   Compute bound sLOGDZGAMMA
403:C   -------------------------
404:        CALL FDZsDILAT(vGAMMA,sRHO1,vT1)
405:        vT1(1)=sONE
406:        CALL FLOG(vT1,vT1)
407:        vT1(1)=sZERO
408:        vT1(2)=sZERO
409:        sLOGDZGAMMA=sQUOT(sCONST(rL1BOUND(vT1)),sPOWER(sRHOP,2))
410:        IF(rMAXABS(sQUOT(sLOGDZGAMMA,sL5)).GT.1.98D+00)WRITE(6,*)
411:     *'error: sLOGDZGAMMA=',sLOGDZGAMMA
412:
413:C   Estimate abs(vDZ(GF0+HF0)): invert Ecalle's equation
414:C   ----------------------------------------------------
415:C..calculate constants
416:        sT1=sEXP(sPROD(sLOGDZGAMMA,sPOWER(sRHOP,2)))
417:        sT2=sINV(sSQRT(sL6))
418:        sT3=sSUM(sQUOT(sLOGDZGAMMA,sL5),sONE)
419:        sT4=sINV(sPROD(sPOWER(sRHOP,2),sL5))
420:C..calculate general
421:        sGREGDZF0=sCONST(rMAXABS(sPROD(sSUM(sONE,sPROD(sT1,sSUM(sONE
422:     *           ,sPROD(sT4,sQUOT(sPOWER(sT2,4),sDIFF(sTHREE,sT3)))))))
423:     *           ,sGDZJTOT)))
424:C..calculate higher order
425:        sHREGDZF0=sCONST(rMAXABS(sPROD(sSUM(sONE,sPROD(sT1,sSUM(sONE
426:     *           ,sPROD(sT4,sQUOT(sPOWER(sT2,N-5)
427:     *           ,sDIFF(sQUOT(siCONST(N-3),sTWO),sT3)))))))
428:     *           ,sHDZJTOT)))
429:
430:C   Collect bounds for sGDZDF0 and sHDZDF0
431:C   --------------------------------------
432:C..collect general
433:        sGDZDF0=sUPPER(sSUM(sGJDZF,sSUM(sPROD(sABS(sDZF01),sRHOP)
434:     *        ,sSUM(sPROD(sABS(sDZF02),sPOWER(sRHOP,2))
435:     *        ,sSUM(sPROD(sABS(sDZF03),sPOWER(sRHOP,3)),sGREGDZF0)))))
436:C..collect higher order
437:        sHDZDF0=sUPPER(sSUM(sHJDZF,sHREGDZF0))
438:
439:
440:C   Calculate ball-dependent parts of COMMON/RES1  /; part 4
441:C   =======================================================
442:
443:C   Calculate functions vDZG00ARG,...
444:C   ---------------------------------
445:        CALL FDZCOMP(vG0,v0ARG,vDZG00ARG)
446:        DO 3 I=1,IBOOT
447:        CALL FDZCOMP(vG0,vARG(1,I),vDZG0ARG(1,I))
448:      3 CONTINUE
449:
450:C   Calculate function vDZhLASTG0
451:C   -----------------------------
452:        CALL FDZCOMP(vhB,vLASTG0,vT1)
453:        CALL FsMULT(vT1,sQUOT(sRHOG0,sRHOhB),vDZhLASTG0)
454:
455:C   Calculate function vDZh
```

```
456:C   -----------------------
457:        CALL FSETINV(vT1,sNEG(sAAA))
458:        CALL FMULT(vT1,vT1,vT1)
459:        CALL FSETLIN(vT2,sNEG(sXOh),sRHOh)
460:        CALL FCOMP(vT1,vT2,vT1)
461:        CALL FsMULT(vT1,sQUOT(sRHOh,sRHOGO),vT1)
462:        CALL FMULT(vT1,vDZGOOARG,vT1)
463:        DO 4 I=1,IBOOT
464:        CALL FMULT(vT1,vDZGOARG(1,I),vT1)
465:      4 CONTINUE
466:        CALL FMULT(vT1,vDZhLASTGO,vDZh)
467:
468:C   Calculate function vDAh
469:C   -----------------------
470:        CALL FEQUAL(vDZh,vT1)
471:        vT1(1)=sZERO
472:        CALL FsMULT(vT1,sPROD(sXOh,sXOh),vT1)
473:        CALL FSETLIN(vT2,sNEG(sXOh),sRHOh)
474:        CALL FMULT(vT2,vT2,vT2)
475:        vT2(1)=sZERO
476:        CALL FMULTADD(vDZh,vT2,vT1)
477:        CALL FsMULT(vT1,sINV(sRHOh),vDAh)
478:
479:C   Calculate function vDTAUG
480:C   -------------------------
481:        CALL FCOMP(vh,vTRANS,vT1)
482:        CALL FsMULT(vT1,sNEG(sINV(sTAU)),vT1)
483:        CALL FCOMP(vDZh,vTRANS,vT2)
484:        CALL FSETLIN(vT3,sSUM(sCENTG,sXOO),sRHOG)
485:        CALL FMULT(vT2,vT3,vT2)
486:        CALL FsMULADD(vT2,sNEG(sQUOT(sLOG(sTAU),sRHOh)),vT1)
487:        CALL FEQUAL(vT1,vDTAUG)
488:
489:C   Calculate function vDAIGOI
490:C   --------------------------
491:        CALL FDZCOMP(vG,vMOEBIUS,vT1)
492:        CALL FSETINV(vT2,sPROD(sAAAI,sRHOGO))
493:        CALL FMULT(vT2,vT2,vT2)
494:        CALL FSHFTR(vT2,vT2)
495:        CALL FSHFTR(vT2,vT2)
496:        CALL FsMULT(vT2,sRHOGO,vT2)
497:        CALL FMULT(vT1,vT2,vDAIGOI)
498:
499:C   Calculate function vACD
500:C   -----------------------
501:        CALL FINVERS(vETA,vT3)
502:        CALL FMULT(vT3,vT3,vT2)
503:        CALL FMULT(vT3,vT2,vT2)
504:        CALL FsMULT(vT2,sPROD(sTWO,sAO),vT1)
505:        CALL FSHFTR(vT3,vT2)
506:        CALL FSHFTR(vT2,vT2)
507:        sT1=sPROD(sRHOGAMMA,sRHOGAMMA)
508:        CALL FsMULADD(vT2,sNEG(sPROD(sPROD(sT1,sTWO),sCO)),vT1)
509:        vT1(4)=sDIFF(vT1(4),sPROD(sPROD(sT1,sRHOGAMMA),sDO))
510:        CALL FsMULT(vT1,sINV(sPROD(sRHOGAMMA,sRHOGAMMA)),vACD)
511:        vACD(2)=sZERO
512:
513:C   Calculate power series xDZFO
514:C   ----------------------------
515:        CALL DEG(N-2)
516:        CALL CDERIVATE(xFO,xT1)
517:        CALL CCOMP(xT1,xGAMMA,xT1)
518:        CALL CSHFTR(xT1,xT1)
519:        CALL FNEG(xT1,xDZFO)
520:
```

```
521:C   Initialize back to degree of vectors
522:C   ------------------------------------
523:        CALL DEG(NSAVE)
524:
525:
526:C   Loop over basis vectors
527:C   =======================
528:
529:C   Prepare loop
530:C   ------------
531:        IF(JPART.EQ.1)THEN
532:          JLOWER=1
533:          JUPPER=6
534:        ELSE
535:          JLOWER=7
536:          JUPPER=30
537:        ENDIF
538:
539:C   Start loop
540:C   ----------
541:        DO 10 JDER=JLOWER,JUPPER
542:
543:C   Generate basis vector
544:C   ---------------------
545:        CALL FSETZERO(vDK0)
546:        vDK0(JDER)=sONE
547:
548:C   Calculate image. Produce output for information
549:C   -----------------------------------------------
550:        CALL FvDKI(vDK0,vDK0I)
551:        rT1=rNORM(vDK0I)
552:        WRITE(6,*)'norm of vector',JDER,rT1
553:
554:C   Take sup of norms
555:C   -----------------
556:        rRHO=rMAX2(rRHO,rNORM(vDK0I))
557:
558:C   End of loop
559:C   -----------
560:     10 CONTINUE
561:
562:
563:C   End of loop JPART 1,2
564:C   =====================
565:
566:     11 CONTINUE
567:
568:
569:C   Take image of higher order ball
570:C   ===============================
571:
572:C   Prepare higher order ball
573:C   -------------------------
574:        CALL FSETZERO(vDK0)
575:        vDK0(NERR)=error(rZERO,rONE)
576:        DO 12 I=JUPPER+1,N1
577:        vDK0(I)=DCMPLX(rZERO,rONE)
578:     12 CONTINUE
579:
580:C   Calculate image. Produce output for information
581:C   -----------------------------------------------
582:        CALL FvDKI(vDK0,vDK0I)
583:        rT1=rNORM(vDK0I)
584:        WRITE(6,*)'norm of higher order',rT1
585:
```

```
586:C  Take sup over norms
587:C  -------------------
588:      rRHO=rMAX2(rRHO,rNORM(vDKOI))
589:
590:
591:C  Produce output for information
592:C  ==============================
593:
594:      WRITE(6,*)´rRHO=´,rRHO
595:
596:
597:C  Check if contraction on ball
598:C  ===========================
599:
600:      IF(rEPSILON.LT.rDOWN(rDOWN(rONE-rRHO)*rBETA))GOTO 55
601:      STOP
602:  55 CONTINUE
603:      WRITE(6,*)´ ´
604:      WRITE(6,*)´        ***********************´
605:      WRITE(6,*)´        * Contraction on ball *´
606:      WRITE(6,*)´        ***********************´
607:      WRITE(6,*)´ ´
608:
609:
610:      END
```

The include-file STUFF

```
    STUFF PROC
          IMPLICIT COMPLEX*16(s,x,v,e)
          IMPLICIT REAL*8(r,a,p)
          PARAMETER NDEG=70
          PARAMETER NHALF=NDEG/2+1
          COMMON/DEGREE/N,N1,N2,NERR
          COMMON/CONST1/sZERO,sONE
          COMMON/CONST2/sTWO,sTHREE,sFOUR,sHALF,s3HALF
          COMMON/CONST3/rZERO,rONE,rTWO,rTHREE,rFOUR,rHALF,r3HALF
          PARAMETER sPI=DCMPLX(3.14159265358979323D+00
        *                    ,3.14159265358979324D+00)
    END
```

The formal structure of this include-file (PROC-FILE) corresponds

to the UNIVAC- conventions.

The include-file STRES1

```
STRES1 PROC
      PARAMETER IBOOT=39
      COMMON/RES1  /sAAA,vGO(NDEG),vGORED(NDEG),vGAMMA(NDEG),vETA(NDEG)
      *              ,vI2ETA(NDEG)
      *              ,vAO(NDEG),vCO(NDEG),vDO(NDEG)
      *              ,xGAMMA(NDEG),sGGAMMA,sHGAMMA
      *              ,vXJ(NDEG)
      *              ,sAO,sCO,sDO,xFO(NDEG),sCFO,sEFO,sC,sR
      *              ,sL5,sL6,sZ,sETA
      *              ,sJOO,sJO1,sJO2,sGDZJ,sHDZJ
      *              ,sGDZDFO,sHDZDFO
      *              ,vhB(NDEG),vh(NDEG)
      *              ,sTAU,sALPHA
      *              ,vTRANS(NDEG),vGPCT(NDEG),vG(NDEG)
      *              ,sAAAI,vMOEBIUS(NDEG)
      *              ,vGOI1(NDEG),vGOI2(NDEG),vGOI(NDEG)
      *              ,vDZGOGO(NDEG)
      *              ,sCQ,sEQ,sCJ,sEJ
      *              ,vACD(NDEG),xDZFO(NDEG)
      *              ,vOARG(NDEG),vARG(NDEG,IBOOT),vLASTGO(NDEG)
      *              ,vDZGOOARG(NDEG),vDZGOARG(NDEG,IBOOT)
      *              ,vDZhLASTGO(NDEG)
      *              ,vDAh(NDEG)
      *              ,vDTAUG(NDEG),vDAIGOI(NDEG)
      END
```

The formal structure of this include-file (PROC-FILE) corresponds
to the UNIVAC-conventions.

We first need to make two general remarks.

a) Input-Output

Computers accept input and produce output in decimal notation. This poses a conceptual problem which is, e.g. illustrated
by the number $1/3$ on a machine with binary internal representation. Thus, if we "give" the number $1/3$ to the machine, it will
actually compute with another number, near to, but not equal to,
one third. We discuss now how this problem is solved. In the case

of <u>input</u>, it is not the input itself for which certain facts hold, but the effective internal representation of this input. For the <u>output</u>, it is not the numbers printed on the sheet of paper, but the content of the corresponding memory registers (which represent a diadic rational). One could use as output the list of bits in this register, and this would be a totally rigorous way of representing reps in output. Such a method is possible, but of course terribly cumbersome. Since <u>output of numbers is for information only</u>, we have chosen to perform decimal output. Thus, when we say that the output has produced a number x, we mean the following : <u>The memory register has contained a rep y, and y differs probably from x at most by one unit in the last figure shown</u>. When we say that, e.g. we work with $\rho_O = 0.726$, $\rho_\Gamma = 0.53,...$ (see Section 5.2), then we mean that the scalars sRHOGØ, sRHOGAMMA,... which represent $\rho_O, \rho_\Gamma,...$ have been generated by the subroutine saCONST (and probably differ from $\rho_O, \rho_\Gamma,...$ at most by one part in 2^{59}). We have checked explicitly that constants in "INITIALIZE" are correctly converted.

All this being said, we henceforth do not readdress this problem.

b)   <u>Showing that a linear map $\ell_1 \to \ell_1$ is a contraction</u>

This is needed as the main ingredient of Theorem 5.4.3.

Identifying A_O with ℓ_1, we can represent $DK_{\{A,G_O\}}$ as a matrix $\Delta \equiv \Delta(\{A,G_O\}) = (\Delta_{ij}(\{A,G_O\}))$, $i,j = 1,...,\infty$, and $\Delta \in \mathcal{L}(\ell_1, \ell_1)$.

We denote $\{\delta A, \delta G_o\}_j$ the element in A_o which is the image, under Δ, of the ℓ_1-sequence $(0, \ldots, e_j=1, 0, \ldots)$. The j-th column of the matrix Δ is then given by

$$(\Delta_{1j}, \Delta_{2j}, \ldots) = DK_{\{A, G_o\}}(\{\delta A, \delta G_o\}_j) .$$

Because we have chosen the ℓ_1-norm, the operator norm of $DK_{\{A, G_o\}}$ is given by

$$\|DK_{\{A, G_o\}}\| = \sup_j \|(\Delta_{1j}, \Delta_{2j}, \ldots)\| . \tag{5.5.1}$$

We now discuss how (5.5.1) can be established on the computer. In principle we need to calculate all the sequences $(\Delta_{1j}, \Delta_{2j}, \ldots)$ $j = 1, 2, \ldots, \infty$. However, inspection of (5.5.1) shows that we only need to estimate the norm of these sequences.

The idea of how to do all these estimations is the following. We treat separately the first $j_{max} < \infty$ image vectors and bound then collectively all cases $j > j_{max}$. This is done as follows. For $1 \le j \le j_{max}$ we fix on the computer the degree K of sequences such that $K > j$, and feed then the standard ball $\{(0, \ldots, e_j=1, 0, \ldots), e^{(G)}=0, e^{(H)}=0\}$ into the computer and calculate (in degree K) the corresponding image vector together with its norm. For $j > j_{max}$, we fix $K > j_{max}$ and feed the standard ball $\{(0, \ldots, e_{j_{max}}=0, [0,1], \ldots, [0,1], v^{(G)}=0, v^{(H)}=1\}$. By definition this ball contains in particular any sequence $\{A, G_o\}_j$ with $j > j_{max}$. Therefore, running the program with this input ball produces as an output an upper bound on $\sup_{j > j_{max}} \|(\Delta_{1j}, \Delta_{2j}, \ldots)\|$, and therefore

(5.5.1) can be estimated in a finite number of arithmetic operations. In the case at hand we fix $j_{max} = 30$ and furthermore we
choose $K = 63$ for $j = 1,...,6$ and $K = 48$ for all $j \geq 7$.
These choices are a compromise between optimizing the computation
time and accuracy of the bounds obtained.

We now discuss briefly the listing of the main program. We
use the letter L to indicate line-numbers. For example $L1-17$
refers to lines 1 to 17 and $L20$ denotes the single line 20.

In a first reading of the program the reader should skip
$L1-22$, $L53-70$, $L98-122$ and $L189-525$. The program is then completely selfexplaining. (In $L126$ we apply the operator K and
calculate then ε. In $L550$ we apply DK. The skipped lines
$L189-525$ contain calculations which we use to evaluate DK, but
which are independent of the column index $JDER \equiv j$. We shall
explain those when we explain the subroutine $FvDKI \equiv DK$.) This
completes our discussion of the main program.

5.6. The subroutine $FvKI \equiv K$

In this section we explain the general structure of the subroutine $FvKI \equiv K$. We shall see that the listing of the program
is again essentially selfexplaining with the exception of the
parts which are concerned with the Borel summation, i.e. the solution of Ecalle's equation. All estimates of this type will be explained in the next section. We give again first the listing of
the program and discuss then its ingredients.

fvki

```
    1:          SUBROUTINE FvKI(vK0,vKOI,IFLAG)
    2:          INCLUDE BIGN.STUFF
    3:          COMMON/DOMAIN/sRHOG0,sRHOGAMMA,sX0h,sCENTh,sRHOh,sCENTG,sRHOG
    4:        *               ,sRHOhB,sRHOPhB,sXOhB,sPHIO,sxKAPPA,sKAPPA
    5:        *               ,sRHOP,sRHO1
    6:          COMMON/SCALE /sX00
    7:          COMMON/SWITCH/ITEST
    8:          COMMON/LOFFEL/sBETA,xPHI(NDEG),vINT1(NHALF,NDEG)
    9:        *             ,vINT2(NHALF,NDEG)
   10:          COMMON/REDUCE/sG03,sG04,sG05,sGA3,sGA33,sGA5,sGA6
   11:          INCLUDE BIGN.STRES1
   12:          DIMENSION vK0(1),vKOI(1)
   13:          DIMENSION vRHS(NDEG),vGMOEBIUS(NDEG)
   14:          DIMENSION xJ(NDEG)
   15:          DIMENSION vT1(NDEG),vT2(NDEG)
   16:
   17:
   18:C UNPACK isomorphism
   19:C =================
   20:
   21:          CALL UNPACK(vK0,sAAA,vG0)
   22:          IF(IFLAG.EQ.1)WRITE(6,*)´sAAA=´,sAAA
   23:
   24:
   25:C Ecalle´s equation: preparatory steps
   26:C ====================================
   27:
   28:C Calculate functions vGORED
   29:C --------------------------
   30:          sT1=sQUOT(sRHOGAMMA,sRHOG0)
   31:          CALL FsDILAT(vG0,sT1,vGORED)
   32:          sT1=sPROD(sCONST(rG(vGORED)),sPOWER(sT1,2))
   33:          vGORED(NERR)=error(rMAXABS(sT1),rH(vGORED))
   34:
   35:C Calculate function vGAMMA
   36:C -------------------------
   37:          CALL FCOMP(vG0,vGORED,vT1)
   38:          sT1=sCONST(rL1NORM(vGORED))
   39:          sT1=sPROD(sCONST(rG(vT1)),sPOWER(sT1,2))
   40:          vT1(NERR)=error(rMAXABS(sT1),rH(vT1))
   41:          CALL FsMULT(vT1,sQUOT(sRHOG0,sRHOGAMMA),vGAMMA)
   42:          vGAMMA(1)=sZERO
   43:          vGAMMA(2)=sONE
   44:          vGAMMA(3)=sZERO
   45:          vGAMMA(5)=sZERO
   46:
   47:C Set constants in COMMON/REDUCE/
   48:C -------------------------------
   49:          sT1=DCMPLX(-rG(vG0),rG(vG0))
   50:          sG03=sSUM(vG0(4),sT1)
   51:          sG04=sSUM(vG0(5),sT1)
   52:          sG05=sSUM(vG0(6),sT1)
   53:          sGA3=sNEG(sQUOT(sPROD(sTWO,sG03),sPOWER(sRHOG0,2)))
   54:          sGA33=sPOWER(sGA3,2)
   55:          sGA5=sQUOT(sDIFF(sPROD(sTHREE,sPOWER(sG03,2)),sPROD(sTWO,sG05))
   56:        *     ,sPOWER(sRHOG0,4))
   57:          sGA6=sNEG(sQUOT(sPROD(sG03,sG04),sPOWER(sRHOG0,5)))
   58:
   59:C Calculate function vETA
   60:C -----------------------
   61:          CALL FSHFTL(vGAMMA,vETA)
   62:          vETA(1)=sONE
   63:
   64:C Calculate function vI2ETA
   65:C -------------------------
```

fvki

```
 66:        CALL FINVERS(vETA,vT1)
 67:        vT1(1)=sONE
 68:        CALL FMULT(vT1,vT1,vI2ETA)
 69:        vI2ETA(1)=sONE
 70:
 71:C   Calculate function vA0
 72:C   ----------------------
 73:        CALL FEQUAL(vI2ETA,vT1)
 74:        vT1(1)=sZERO
 75:        CALL FSHFTL(vT1,vT1)
 76:        vT1(1)=sZERO
 77:        CALL FSHFTL(vT1,vT1)
 78:        CALL FsMULT(vT1,sINV(sPROD(sRHOGAMMA,sRHOGAMMA)),vA0)
 79:C..more precise value
 80:        vA0(1)=sNEG(sPROD(sTWO,sGA3))
 81:
 82:C   Calculate function vC0
 83:C   ----------------------
 84:        CALL FLOG(vETA,vC0)
 85:
 86:C   Calculate function vD0
 87:C   ----------------------
 88:        CALL FEQUAL(vGAMMA,vT1)
 89:        vT1(2)=sZERO
 90:        CALL FsMULT(vT1,sRHOGAMMA,vD0)
 91:
 92:
 93:C   Prove that Ecalle's equation has analytic solution on prescribed dom
 94:C   ===================================================================
 95:
 96:C   1) Check domain in t-plane
 97:C   ==========================
 98:
 99:C   Calculate sKAPPA
100:C   ----------------
101:        sKAPPA=sNEG(sQUOT(sPROD(sTWO,sPROD(sRHOGAMMA,sRHOGAMMA))
102:     *        ,sSUM(vGAMMA(4),DCMPLX(-rG(vETA),rG(vETA)))))
103:        IF(IFLAG.EQ.1)WRITE(6,*)'sKAPPA=',sKAPPA
104:
105:C   Check sBETA
106:C   -----------
107:        IF(rMAXABS(sBETA).GT.rMINABS(sQUOT(sKAPPA,sTWO)))THEN
108:          WRITE(6,*)'error in choice of sBETA',sBETA
109:        ENDIF
110:
111:
112:C   2) Check domains in z-plane
113:C   ===========================
114:
115:C   Map domain of definition of vh into z-plane of normal forms
116:C   -----------------------------------------------------------
117:C..calculate vOARG
118:        sT1=sSUM(sONE,sPROD(sAAA,sXOh))
119:        CALL FSETINV(vT1,sPROD(sQUOT(sRHOh,sXOh)
120:     *             ,sDIFF(sINV(sT1),sONE)))
121:        CALL FsMULT(vT1,sINV(sT1),vT1)
122:        vT1(1)=sDIFF(vT1(1),sONE)
123:        CALL FsMULT(vT1,sINV(sPROD(sAAA,sRHOGO)),vOARG)
124:C..more precise constant term
125:        vOARG(1)=sNEG(sQUOT(sXOh,sPROD(sRHOGO,sT1)))
126:
127:C   Check that this disc is contained in sector S(rhogamma,phi0)
128:C   -----------------------------------------------------------
129:        sT1=sNEG(sPROD(vOARG(1),sRHOGO))
130:        CALL FEQUAL(vOARG,vT1)
```

fvki

```
131:        vT1(1)=sZERO
132:        sT2=sPROD(sCONST(rL1NORM(vT1)),sRHOGO)
133:C..check radius
134:        rT1=rMAXABS(sSUM(sT1,sT2))
135:        rT2=sRHOGAMMA
136:        IF(rT1.GT.rT2)WRITE(6,*)'error in radius 1)',rT1,rT2
137:C..check angle
138:        rT1=rMAXABS(sQUOT(sT2,sT1))
139:        rT2=rMINABS(sSIN(sPHIO))
140:        IF(rT1.GT.rT2)WRITE(6,*)'error in angle 1)',rT1,rT2
141:
142:C   Check that the domain of definition of vhB is contained in sector
143:C   ----------------------------------------------------------------
144:C..check radius
145:        rT1=rMAXABS(sSUM(sXOhB,sRHOPhB))
146:        rT2=sRHOGAMMA
147:        IF(rT1.GT.rT2)WRITE(6,*)'error in radius 2)',rT1,rT2
148:C..check angle
149:        rT1=rMAXABS(sQUOT(sRHOPhB,sXOhB))
150:        rT2=rMINABS(sSIN(sPHIO))
151:        IF(rT1.GT.rT2)WRITE(6,*)'error in angle 2)',rT1,rT2
152:
153:C   Check that sector is invariant
154:C   -----------------------------
155:        sRHOTEST=sPROD(sRHOGAMMA,saCONST(.74D+00))
156:        CALL FsDILAT(vGO,sQUOT(sRHOTEST,sRHOGO),vT1)
157:        vT1(2)=sZERO
158:        vT1(4)=sZERO
159:        rT1=rMAXABS(sQUOT(sPROD(sCONST(rL1BOUND(vT1)),sRHOGO)
160:     *     ,sPOWER(sRHOTEST,3)))
161:        sX=sQUOT(sSUM(sRHOGAMMA,sRHOTEST),sPROD(sTWO,sCOS(sPHIO)))
162:        sR=sUPPER(sSQRT(sDIFF(sSUM(sPOWER(sX,2),sPOWER(sRHOTEST,2))
163:     *     ,sPROD(sTWO,sPROD(sPROD(sRHOTEST,sX),sCOS(sPHIO))))))
164:        CALL FSHFTL(vGO,vT1)
165:        vT1(1)=sZERO
166:        CALL FSHFTL(vT1,vT1)
167:        vT1(1)=sZERO
168:        CALL FSHFTL(vT1,vT1)
169:        vT1(1)=sZERO
170:        CALL FSETLIN(vT2,sQUOT(sX,sRHOGO),sQUOT(sR,sRHOGO))
171:        CALL FCOMP(vT1,vT2,vT2)
172:        rT2=rMAXABS(sQUOT(sCONST(rL1NORM(vT2)),sPOWER(sRHOGO,2)))
173:        CALL FSETLIN(vT2,sNEG(sQUOT(sX,sRHOGO)),sQUOT(sR,sRHOGO))
174:        CALL FCOMP(vT1,vT2,vT2)
175:        rT3=rMAXABS(sQUOT(sCONST(rL1NORM(vT2)),sPOWER(sRHOGO,2)))
176:        sT1=sCONST(rMAX3(rT1,rT2,rT3))
177:C..check contraction of length
178:        rT1=rMAXABS(sQUOT(sPOWER(sRHOGAMMA,2),sKAPPA))
179:        IF(rT1.GT.rONE)WRITE(6,*)'error in invariance check 1)',rT1
180:        rT1=rMAXABS(sT1)
181:        rT2=rMINABS(sQUOT(sDIFF(sCOS(sPROD(sTWO,sPHIO))
182:     *     ,sSIN(sPROD(sTWO,sPHIO))),sKAPPA))
183:        IF(rT1.GT.rT2)WRITE(6,*)'error in invariance check 2)',rT1,rT2
184:C..check contraction of angles
185:        rT1=rMAXABS(sPROD(sTHREE,sQUOT(sPOWER(sRHOGAMMA,2),sKAPPA)))
186:        IF(rT1.GT.rONE)WRITE(6,*)'error in invariance check 3)',rT1
187:        sT3=sPOWER(sQUOT(sSIN(sPHIO),sCOS(sPHIO)),2)
188:        sT4=sSQRT(sSUM(sONE,sT3))
189:        sT5=sSUM(sQUOT(sPOWER(sRHOGAMMA,2),sKAPPA),sPROD(sPROD
190:     *     (sPOWER(sRHOGAMMA,2),sPOWER(sT4,3)),sT1))
191:        rT1=rMAXABS(sT5)
192:        IF(rT1.GT.rONE)WRITE(6,*)'error in invariance check 4)',rT1
193:        CALL FDZsDILAT(vT1,sQUOT(sRHOTEST,sRHOGO),vT2)
194:        rT1=rMAXABS(sPROD(sRHOTEST,sCONST(rL1BOUND(vT2))))
195:        sTRHO=sSQRT(sQUOT(sRHOGAMMA,sRHOGO))
```

fvki

```
196:        CALL FDZsDILAT(vT1,sRHO,vT1)
197:        CALL FSHFTR(vT1,vT1)
198:        CALL FsMULT(vT1,sRHO,vT1)
199:        sT5=sSQRT(sPROD(sRHOGO,sRHOGAMMA))
200:        CALL FSETLIN(vT2,sQUOT(sX,sT5),sQUOT(sR,sT5))
201:        CALL FCOMP(vT1,vT2,vT2)
202:        rT2=rL1NORM(vT2)
203:        CALL FSETLIN(vT2,sQUOT(sX,sT5),sNEG(sQUOT(sR,sT5)))
204:        CALL FCOMP(vT1,vT2,vT2)
205:        rT3=rL1NORM(vT2)
206:        sT2=sQUOT(sCONST(rMAX3(rT1,rT2,rT3)),sPOWER(sRHOGO,3))
207:        sT2=sSUM(sPROD(sTHREE,sT1),sT2)
208:        sT5=sSUM(sT3,sSUM(sPROD(sKAPPA
209:      *      ,sPROD(sPOWER(sT4,3),sT1)),sPROD(sKAPPA
210:      *      ,sPROD(sPOWER(sT4,2),sT2))))
211:        rT1=rMAXABS(sT5)
212:        IF(rT1.GT.rTWO)WRITE(6,*)´error in invariance check 5)´,rT1
213:
214:C   Prove that sector is attracted by Z=0
215:C   -----------------------------------
216:        CALL FEQUAL(vI2ETA,vT1)
217:        vT1(1)=sZERO
218:        vT1(2)=sZERO
219:        sT1=vT1(3)
220:        vT1(3)=sZERO
221:        rT1=rMINABS(sDIFF(sT1,sCONST(rL1NORM(vT1))))
222:        IF(rT1.LT..0001D+00)WRITE(6,*)´error:domain is not attracted´,rT1
223:
224:C   Calculate bounds of COMMON/RES1/
225:C   -------------------------------
226:C..calculate sL5
227:        CALL FsDILAT(vI2ETA,sRHO1,vT1)
228:        CALL FsMULT(vT1,sINV(sPROD(sRHOP,sRHOP)),vT1)
229:        vT1(1)=sZERO
230:        sT1=vT1(3)
231:        vT1(3)=sZERO
232:        sL5=sLOWER(sDIFF(sT1,sCONST(rL1NORM(vT1))))
233:C..calculate sL6 and sL7
234:        sL6=sCOS(sPROD(sTWO,sPHIO))
235:        sL7=sSIN(sPROD(sTWO,sPHIO))
236:C..calculate sZ
237:        CALL FsDILAT(vETA,sRHO1,vT1)
238:        CALL FMULT(vT1,vT1,vT1)
239:        vT1(1)=sZERO
240:        sT1=sCONST(rL1BOUND(vT1))
241:        sZ=sUPPER(sINV(sDIFF(sPROD(sL6,sDIFF(sONE,sT1)),sPROD(sL7,sT1))))
242:C..calculate sETA
243:        CALL FsDILAT(vI2ETA,sRHO1,vT1)
244:        sETA=sINV(sSQRT(sCONST(rL1NORM(vT1))))
245:C..calculate bounds sGGAMMA and sHGAMMA
246:        sGGAMMA=sPROD(sPROD(sRHOGAMMA,sCONST(rG(vGAMMA)))
247:      *          ,sPOWER(sRHO1,3))
248:        sHGAMMA=sPROD(sPROD(sRHOGAMMA,sCONST(rH(vGAMMA)))
249:      *          ,sPOWER(sRHO1,N1))
250:
251:
252:C   Ecalle´s equation: calculate function vh
253:C   ========================================
254:
255:C   Produce inhomogeneous part of Ecalle´s equation
256:C   -----------------------------------------------
257:        CALL FSETZERO(vRHS)
258:        vRHS(1)=sNEG(sTWO)
259:        CALL CREDUCE(vRHS,vAO,vCO,vDO,sAO,sCO,sDO,vxJ,xJ,sGJ,sHJ
260:      *              ,IFLAG,sJOO,sJO1,sJO2,sGDZJ,sHDZJ)
```

fvki

```
261:      *                ,vI2ETA,sCFO,sEFO)
262:         IF(IFLAG.EQ.1)WRITE(6,*)´sAO´,sAO
263:         IF(IFLAG.EQ.1)WRITE(6,*)´sCO´,sCO
264:         IF(IFLAG.EQ.1)WRITE(6,*)´sDO´,sDO
265:
266:C   Produce power series solution of Ecalle´s equation
267:C   --------------------------------------------------
268:         CALL CxFO(xGAMMA,xJ,xFO,N)
269:
270:
271:C   Remark: do not recalculate integral table in run for derivative.
272:C   ------
273:         IF(IFLAG.EQ.1)GOTO 111
274:
275:C   Prepare functions for Borel summation
276:C   -------------------------------------
277:         CALL PREPBOREL(sXOhB,sRHOhB,N)
278:
279:
280:C   Remark: Borel resummation starts here
281:C   ------
282:   111 CONTINUE
283:
284:
285:C   Calculate Loeffel/Borel-bounds
286:C   ------------------------------
287:         sC=sUPPER(sPROD(sCFO,sQUOT(sFOUR,sxKAPPA)))
288:         sR=sLOWER(sINV(sPROD(sEFO,sQUOT(sFOUR,sxKAPPA))))
289:
290:C   Sum the formal power series part of FO to the sectorial solution
291:C   ---------------------------------------------------------------
292:         CALL FSUMBOREL(xFO,N,sC,sR,vhB)
293:
294:C   Add contribution of general term
295:C   --------------------------------
296:         CALL FADDGENERAL(vhB,sGJ,sHJ,sC,sR,sGGAMMA,sHGAMMA
297:      *                  ,sZ,sETA,sL5,sL6,vhB)
298:
299:C   Restore irregular inhomogeneous part
300:C   ------------------------------------
301:         CALL FCOMPLETE(vhB,sAO,sCO,sDO,vhB)
302:
303:C   Analytic bootstrap
304:C   ------------------
305:         CALL FCOMP(vGO,vOARG,vARG(1,1))
306:         DO 1 I=2,IBOOT
307:         CALL FCOMP(vGO,vARG(1,I-1),vARG(1,I))
308:     1 CONTINUE
309:         CALL FCOMP(vGO,vARG(1,IBOOT),vT1)
310:         CALL FsMULT(vT1,sQUOT(sRHOGO,sRHOhB),vT1)
311:         vT1(1)=sSUM(vT1(1),sQUOT(sXOhB,sRHOhB))
312:         CALL FEQUAL(vT1,vLASTGO)
313:         CALL FCOMP(vhB,vLASTGO,vh)
314:         vh(1)=sZERO
315:
316:
317:C   Determine sTAU
318:C   ==============
319:
320:         sTAU=sFvZERO(vh)
321:         IF(IFLAG.EQ.1)WRITE(6,*)´sTAU=´,sTAU
322:
323:
324:C   Determine sALPHA
325:C   ================
```

fvki

```
326:
327:        sT1=sQUOT(sDIFF(sPROD(sTAU,sX00),sCENTh),sRHOh)
328:        sALPHA=sPROD(sX00,sEXP(sPROD(sLOG(sTAU),sVALUE(vh,sT1))))
329:
330:
331:C  Determine sX0 (for information only)
332:C  ====================================
333:
334:        IF(IFLAG.EQ.0)GOTO 3
335:        sX0=sPROD(sQUOT(sX00,sALPHA),sEXP(sPROD(sLOG(sTAU),sVALUE(vh
336:      *         ,sNEG(sQUOT(sCENTh,sRHOh))))))
337:        WRITE(6,*)'sX0=',sX0
338:      3 CONTINUE
339:
340:
341:C  Calculate vector vG
342:C  ===================
343:
344:C  Translate vh to domain of vG
345:C  ----------------------------
346:        CALL FSETLIN(vTRANS,sQUOT(sDIFF(sPROD(sTAU,sSUM(sCENTG,sX00))
347:      *            ,sCENTh),sRHOh),sQUOT(sPROD(sTAU,sRHOG),sRHOh))
348:        CALL FCOMP(vh,vTRANS,vT1)
349:
350:C  Multiply with -log(sTAU)
351:C  ------------------------
352:        CALL FsMULT(vT1,sNEG(sLOG(sTAU)),vT1)
353:
354:C  Take exponential of non-constant part
355:C  -------------------------------------
356:        sT1=vT1(1)
357:        vT1(1)=sZERO
358:        CALL FEXP(vT1,vT1)
359:
360:C  Restore constant part and normalize
361:C  -----------------------------------
362:        CALL FsMULT(vT1,sQUOT(sPROD(sEXP(sT1),sALPHA),sRHOG),vGPCT)
363:
364:C  Subtract constant term and store result in vG
365:C  ---------------------------------------------
366:        CALL FEQUAL(vGPCT,vT1)
367:        vT1(1)=sDIFF(vT1(1),sQUOT(sSUM(sX00,sCENTG),sRHOG))
368:        CALL FEQUAL(vT1,vG)
369:
370:
371:C  Calculate sAAAI
372:C  ===============
373:
374:        sT1=sQUOT(sDIFF(sPROD(sTAU,sX00),sCENTh),sRHOh)
375:        sT2=sQUOT(sTAU,sRHOh)
376:        sT2=sPROD(sT2,sT2)
377:        sAAAI=sPROD(sNEG(sINV(sFOUR)),sDIFF(sINV(sX00),sPROD(sX00
378:      *        ,sPROD(sLOG(sTAU),sPROD(sT2,sDVALUE(2,vh,sT1))))))
379:
380:
381:C  Conjugate vG to obtain vGOI
382:C  ===========================
383:
384:C  Construct Moebius mapping and normalize it to unit disc
385:C  -------------------------------------------------------
386:        CALL FSETINV(vT1,sPROD(sAAAI,sRHOGO))
387:        CALL FSHFTR(vT1,vT1)
388:        CALL FsMULT(vT1,sQUOT(sRHOGO,sRHOG),vT1)
389:        vT1(1)=sDIFF(vT1(1),sQUOT(sCENTG,sRHOG))
390:        CALL FEQUAL(vT1,vMOEBIUS)
```

```
391:
392:
393:C  Substitute vMOEBIUS
394:C  -------------------
395:       CALL FCOMP(vG,vMOEBIUS,vGMOEBIUS)
396:
397:C  Construct vGOI1,vGOI2 and vGOI and normalize properly
398:C  -----------------------------------------------------
399:       CALL FsMULT(vGMOEBIUS,sQUOT(sRHOG,sRHOGO),vT1)
400:       vT1(1)=sSUM(vT1(1),sQUOT(sCENTG,sRHOGO))
401:       CALL FEQUAL(vT1,vGOI1)
402:C..more precise values
403:       vGOI1(1)=sZERO
404:       vGOI1(2)=sNEG(sONE)
405:       CALL FsMULT(vGMOEBIUS,sNEG(sPROD(sAAAI,sRHOG)),vT1)
406:       vT1(1)=sONE
407:       CALL FINVERS(vT1,vGOI2)
408:C..more precise values
409:       vGOI2(1)=sONE
410:       vGOI2(3)=sZERO
411:       CALL FMULT(vGOI1,vGOI2,vGOI)
412:       vGOI(1)=sZERO
413:       vGOI(2)=sNEG(sONE)
414:       vGOI(3)=sZERO
415:
416:
417:C  PACK isomorphism
418:C  ================
419:
420:       CALL PACK(sAAAI,vGOI,vKOI)
421:
422:
423:       RETURN
424:       END
```

(Note that IFLAG = 0 in the subroutine call for the case where the subroutine means the application of K. The case IFLAG = 1 is treated in the discussion of the subroutine FvDKI $\equiv$ DK. See below.)

The discussion of this subroutine can essentially be reduced to write all equations defining the operator K ((5.3.1)-(5.3.10)) for vectors instead of functions. This is exactly what our subroutine package which has been explained in Section 4 does in an automatic fashion except for the construction of the function h_B^u of (5.3.3). This construction will be put aside for the time being (L47-114 and L126-303 of the subroutine FvKI). For the remaining lines we have :

i) Comments on (5.3.1). Given $vK\emptyset \equiv \{A, G_0\} \epsilon A_1$ we construct

successively $sAAA \equiv A$ and $vG\emptyset \equiv G_0^u(z)$, $vG\emptyset RED \equiv G_0^u(\frac{\rho_\Gamma}{\rho_0}z)$,

and $vGAMMA \equiv \Gamma^u(z)$. (See L21, L30-33, and L37-45 res-

pectively.)

ii) (5.3.2) and (5.3.3) are put aside and we continue by sup-

posing that the function h_B^u is already known. Comments on

(5.3.4): We construct $v\emptyset ARG \equiv \frac{1}{\rho_0}T_{-A}(-x_{oh} + \rho_h z)$ in

L115-125. Then in L305-308 we successively calculate the

iterates $vARG[\nu] \equiv G_0^\nu(\frac{1}{\rho_0}T_{-A}(-x_{oh} + \rho_h z))$, $\nu = 1,\ldots,IBOOT$,

and finally $vLASTG\emptyset \equiv G_0^{IBOOT+1}(\frac{1}{\rho_0}T_{-A}(-x_{oh} + \rho_h z))$ in

L309-312. Substituting in h_B^u in L313 leads to the func-

tion h^u, up to a trivial additive constant which is such

that $h^u(0) = 0$ (L314).

iii) Comment on (5.3.5). To determine τ we use a variant of

Newton algorithm as explained in Section 4 in the case of

the functions sLOG and sSQRT. This is done in the func-

tion sFvZERO in L320. We give the listing of this func-

tion which is completely selfexplaining after what has been

explained in Section 4.

```
      COMPLEX FUNCTION sFvZERO*16(v)
C..computes tau from d/dz exp(-log tau*v(tau*z)) (at z=x00) =-1
C   i.e. 1-v´(tau*x00)*tau*x00*log tau =0
      INCLUDE BIGN.STUFF
      COMMON/DOMAIN/sRHOGO,sRHOGAMMA,sXOh,sCENTh,sRHOh,sCENTG,sRHOG
     *             ,sRHOhB,sRHOPhB,sXOhB,sPHIO,sxKAPPA,sKAPPA
     *             ,sRHOP,sRHO1
      COMMON/SCALE /sx00
      COMMON/SWITCH/ITEST
      DIMENSION v(1)
      DIMENSION p(NDEG)
```

```
C..initial guess
      CALL AvTOp(v,p)
      aX00=aCENT(sX00)
      aCENTh=aCENT(sCENTh)
      aRHOh =aCENT(sRHOh)
      a=.0333
      DO 1 I=1,20
      aARG=(a*aX00-aCENTh)/aRHOh
      aV1=aDXVALUE(p,aARG)
      aN=a*aX00/aRHOh
      aT1=rONE-aN*DLOG(a)*aV1
      aT2=-(aX00/aRHOh)*((rONE+DLOG(a))*aV1+aN*DLOG(a)
     *        *aDXXVALUE(p,aARG))
      a=a-aT1/aT2
    1 CONTINUE
      s=saCONST(a)
C..s is approx root
C..compute values at central guess
      sARG=sQUOT(sDIFF(sPROD(s,sX00),sCENTh),sRHOh)
      sD1=sQUOT(sDVALUE(1,v,sARG),sRHOh)
      sN=sPROD(s,sX00)
      s1=sDIFF(sONE,sPROD(sN,sPROD(sLOG(s),sD1)))
      sEPSILON=s1
C..we increase epsilon to contain 0
      sEPSILON=DCMPLX(rMIN2(DREAL(sEPSILON),rZERO)
     *               ,rMAX2(DIMAG(sEPSILON),rZERO))
C..Newton for intervals
      rBETA=1.D-5
      sBETA=DCMPLX(-rBETA,rBETA)
      DO 2 I=1,10
      sINT=sSUM(s,sBETA)
      sARG=sQUOT(sDIFF(sPROD(sINT,sX00),sCENTh),sRHOh)
      sD1=sQUOT(sDVALUE(1,v,sARG),sRHOh)
      sD2=sQUOT(sDVALUE(2,v,sARG), sPROD(sRHOh,sRHOh))
      sN=sPROD(sINT,sX00)
      sDER=sNEG(sPROD(sX00,sSUM(sPROD(sSUM(sONE,sLOG(sINT)),sD1)
     *        ,sPROD(sN,sPROD(sLOG(sINT),sD2)) )))
      sNEW=sNEG(sQUOT(sEPSILON,sDER))
      IF(sINTER(sNEW,sBETA).NE.sNEW)THEN
        WRITE(6,*)´error in sFvZERO, no contraction´,sNEW
        sFvZERO=sINT
        RETURN
      ENDIF
      sBETA=sNEW
    2 CONTINUE
      sFvZERO=sSUM(s,sBETA)
      RETURN
      END
```

iv) Comments on (5.3.6)-(5.3.10) (L324-420). These lines are

selfexplanatory. We have the identifications $sALPHA \equiv \alpha$,

$sx\emptyset \equiv x_o$,

$$\text{vTRANS} \equiv \frac{\tau \rho_G}{\rho_h} z + \frac{\tau(\alpha_G + x_{oo}) - \alpha_h}{\rho_h} \ ,$$

$$\text{vGPCT} \equiv \widetilde{G}^u(z) + \frac{x_{oo} + \alpha_G}{\rho_G} \ , \qquad \text{vG} \equiv \widetilde{G}^u(z) \ ,$$

$$\text{sAAAI} \equiv \widetilde{A} \ ,$$

$$\text{vMOEBIUS} \equiv \frac{1}{\rho_G}(T_{\widetilde{A}}(\rho_o z) - \alpha_G) \ ,$$

$$\text{vGMOEBIUS} \equiv \widetilde{G}^u(\frac{1}{\rho_G} T_{\widetilde{A}}(\rho_o z) - \alpha_G) \ ,$$

$$\text{vGOI1} \equiv \frac{\alpha_G}{\rho_o} + \frac{\rho_G}{\rho_o} \cdot \widetilde{G}^u(\frac{1}{\rho_G} T_{\widetilde{A}}(\rho_o z) - \alpha_G) \ ,$$

$$\text{vGOI2} \equiv [1 - A\alpha_G - A\rho_G \widetilde{G}^u(\frac{1}{\rho_G} T_{\widetilde{A}}(\rho_o z) - \alpha_G)]^{-1} \ ,$$

$$\text{vGOI} \equiv \widetilde{G}_o^u(z) \quad \text{(L411)} \ , \quad \text{and finally}$$

$$\text{vKOI} \equiv \{\widetilde{A}, \widetilde{G}_o\} \epsilon A_1 \ .$$

This completes our discussion of the subroutines FvKI with the exception of all those parts connected to the solution of Ecalle's equation. This is the content of the next section.

5.7. The constructive solution of Ecalle's equation

This is a very essential section of our paper. Here, we explain how to solve Ecalle's equation constructively on a computer. In fact, we solve several different problems in this subsection.

1) Given polynomials Γ_P and J_P , we construct from the formal power series solution, a solution (i.e. a "vector") of the equation

$$F(\Gamma_P) - F = J_P .$$

Here we use a computer adapted version of Loeffel's method (Section 5.7.4).

2) Given a function (i.e. a "vector") $\Gamma = \Gamma_P + \Gamma_G + \Gamma_H$ and an inhomogeneity ΔJ, we construct a vector ΔF which solves

$$\Delta F(\Gamma) - \Delta F = \Delta J .$$

ΔF will only have general and higher order terms, and we only need to assume bounds on ΔJ. (Section 5.7.5.)

3) Given Γ, and F as constructed in 1), we are able to give bounds on $F(\Gamma) - F$. This will be useful for producing bounds on the r.h.s. as in 2). (Section 5.7.5.)

A more detailed description will be given in Table 5.7.1, below.

5.7.1 Domain questions

An important first step is to verify the compatibility of the analyticity domains which we have chosen. This is done in L93-249 of the subroutine FvKI. We discuss now these lines in details.

In the t-plane (Borel plane of normal forms as described in Section 2) we want to bound functions in a domain

$S_{2,\tilde{\beta}} = \{-\tilde{\beta}\log(1-t) \,|\, t\in\mathbb{C}_2, |t|<1\}$, and we choose $\tilde{\beta} = 2$. See Fig. 5.7.1

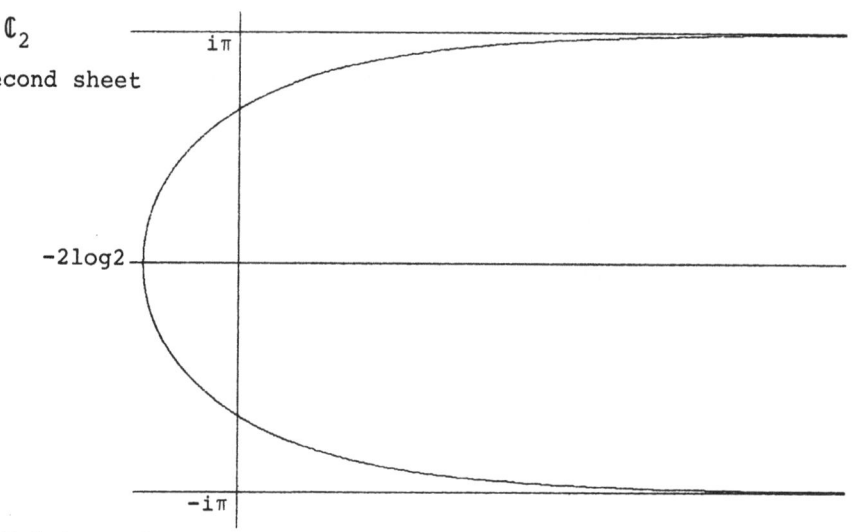

$\mathbb{C}_2$

second sheet

$i\pi$

$-2\log 2$

$-i\pi$

Fig. 5.7.1. Analyticity domain in Ecalle's Borel plane.

In this domain we shall need bounds on the functions

$$E_m(t) = \frac{(-t)^m/m!}{(e^t-1)}, \quad m\in\mathbb{N} ,$$

and on

$$E_\alpha(t) = \frac{1-e^{-\alpha|t|}}{1-e^{-t}}, \quad \alpha \geq 1 ,$$

Lemma 5.7.1. **Let** $S_{2,2} = \{-2\log(1-t)\,|\,t\in\mathbb{C}_2, |t|<1\}$. **Then**

$$\sup_{t\in S_{2,2}} |E_m(t)| \leq C_3 \tag{5.7.1}$$

$$\sup_{t \in S_{2,2}} |E_\alpha(t)| \leq 1 + C_3(1 + \alpha) , \qquad\qquad (5.7.2)$$

where $C_3 = -E_1(-2\log 2) = (3 \log 2)/8 < 1.8483999.$

The proof of (5.7.1) is trivial but lengthy and is left to the reader. We get (5.7.2) as follows

$$\left| \frac{1-e^{-\alpha|t|}}{1-e^{-t}} \right| \leq 1 + \left| \frac{1-e^{-\alpha|t|}e^{t}}{e^{t}-1} \right|$$

$$= 1 + \left| \frac{t}{e^{t}-1} \right| \cdot \left| \frac{1-e^{-\alpha|t|}e^{t}}{t} \right|$$

$$\leq 1 + C_3 \cdot \left| \frac{1-e^{-\alpha|t|}-e^{-\alpha|t|}(e^{t}-1)}{t} \right|$$

$$\leq 1 + C_3 \left\{ \left| \frac{1-e^{-\alpha|t|}}{t} \right| + e^{-\alpha|t|} \left| \frac{e^{t}-1}{t} \right| \right\} .$$

We have

$$\left| \frac{1-e^{-\alpha|t|}}{t} \right| = \frac{1}{|t|} \int_{-\alpha|t|}^{o} e^{\xi}d\xi \leq \frac{1}{|t|} \int_{-\alpha|t|}^{o} d\xi = \alpha ,$$

and

$$\left| \frac{e^{t}-1}{t} \right| \leq \sum_{n=1}^{\infty} \frac{|t|^{n-1}}{n!} \leq \sum_{n=1}^{\infty} \frac{|t|^{n-1}}{(n-1)!} = e^{|t|} ,$$

and therefore

$$|E_\alpha(t)| \leq 1 + C_3 \cdot \{\alpha + e^{-(\alpha-1)|t|}\} \leq 1 + C_3\{\alpha+1\} ,$$

where the last inequality is true because $\alpha \geq 1$ by assumption.

Remark : The constant $C_3 = sC3$ is fixed in L61 of the main program.

We explain now L99-109. Let again g be the function of normal form which is associated to Γ of (5.3.2). Let furthermore f be the function associated to g by Theorem 2.4.4 and f_B its Borel transform. We have

$$f(z) = \int_0^\infty e^{-t/z} f_B(t)\, dt \ .$$

Define now the function $F(z) = f(\frac{4}{\kappa}z^2)$, where $\kappa = \frac{-12}{\Gamma'''(0)}$. Then

$$F(z) = \int_0^\infty e^{-t/z^2} (\frac{4}{\kappa^2} f_B(\frac{4}{\kappa}t))\, dt \ . \tag{5.7.3}$$

We define $f_{B,\mathcal{L}}(t) = \frac{4}{\kappa^2} f_B(\frac{4}{\kappa}t)$. We shall only work with the restriction of f_B to $S_{2,2}$. $f_{B,\mathcal{L}}$ is therefore analytic on $S_{2,\beta}$, where $\beta = \frac{\kappa}{4} \cdot 2 = \frac{\kappa}{2}$. Furthermore if f_B satisfies the bound $|f_B(t)| \leq C_f \exp(E_f|t|)$ for $t\in S_{2,2}$, then $f_{B,\mathcal{L}}$ satisfies

$$|f_{B,\mathcal{L}}(t)| \leq C_{\mathcal{L}}\exp(|t|/R_{\mathcal{L}}) \ , \quad t\in S_\beta \ ,$$

with $C_{\mathcal{L}} = \frac{4}{\kappa^2} C_f$, $R_{\mathcal{L}} = (\frac{4}{\kappa}E_f)^{-1}$. We shall apply the constructive method of Loeffel using $f_{B,\mathcal{L}}$ instead of f_B , the reason being that the coefficients of $f_{B,\mathcal{L}}$ can be calculated with greater precision from the coefficients of Γ. This is essential, since the algorithm which permits to calculate these coefficients with rigorous up and down rounding is unstable. Again to gain preci-

sion we do not calculate β from κ, but <u>choose</u>

sBETA $\equiv \beta$ = .9065 (L63 in the main program) and verify in

L107 that for this choice of β the inequality $\beta < \frac{\kappa}{2}$ is sat-

isfied, where sKAPPA $\equiv \kappa$ is calculated in L101-102 (using

that $\Gamma^u(z) = z - \frac{2}{\kappa}(\rho_\Gamma)^2 z^3 + O(z^4)$).

We next verify the geometric relation between the domains

discussed in Section 5.3. We verify all the necessary bounds

directly in the z-plane. (The corresponding geometric relations

in $\mathbb{C}_2$ are an evident transcription of the ones in the z-plane.)

We shall need for example to verify that $G_o(S^\pm) \subset S^\mp$, for

the functions G_o at hand and $S^\pm$ as defined in (5.2.5). To

achieve this we shall need bounds on the supremum of functions

which are analytic on $S^\pm$, i.e. we need to establish, say, a

constant c such that for a given function f, $\sup\limits_{z \in S^\pm} |f(z)| \leq c$.

Take as an example the function G_o, then

$$\sup_{z \in S^\pm} |G_o(z)| \leq \sup_{z \in D(0,\rho_\Gamma)} |G_o(z)| = \left(\sup_{z \in D(0,1)} |G_o^u(\frac{\rho_\Gamma}{\rho_o}z)| \right) \cdot \rho_o$$

$$\leq |G_o^u(\frac{\rho_\Gamma}{\rho_o}z)|_1 \cdot \rho_o = \text{rL1NORM(vGØRED)} \cdot \rho_o \ .$$

Unfortunately, bounds which are established this way are very

often too poor for our needs, but can be sharpened by considering

the relative sign of the first few Taylor coefficients of the

function in question, when developed at $z = 0$. This is achieved

by the following subroutine, called rL1BOUND.

```
      REAL FUNCTION rL1BOUND*8(v)
      INCLUDE BIGN.STUFF
      COMMON/DOMAIN/sRHOGO,sRHOGAMMA,sXOh,sCENTh,sRHOh,sCENTG,sRHOG
     *            ,sRHOhB,sRHOPhB,sXOhB,sPHIO,sxKAPPA,sKAPPA
     *            ,sRHOP,sRHO1
      DIMENSION v(1)
      DIMENSION vT1(NDEG)
      CALL FEQUAL(v,vT1)
      I=1
    1 CONTINUE
      rT1=rMAXABS(vT1(I))
      IF(rT1.EQ.rZERO.AND.I.LT.N-1)THEN
        I=I+1
        GOTO 1
      ENDIF
      sT1=sPROD(vT1(I),vT1(I+2))
      IF(DIMAG(sT1).GE.rZERO.OR.I.EQ.N-1)THEN
        rL1BOUND=rL1NORM(vT1)
      ELSE
        rT1=rMAXABS(vT1(I))
        rT2=rMAXABS(sSQRT(sSUM(sPOWER(sSUM(vT1(I),sPROD(vT1(I+2)
     *      ,sCOS(sPROD(sTWO,sPHIO)))),2),sPOWER(sPROD(vT1(I+2)
     *      ,sSIN(sPROD(sTWO,sPHIO))),2))))
        rT3=rMAXABS(sSUM(vT1(I),vT1(I+2)))
        vT1(I)=sZERO
        vT1(I+2)=sZERO
        rL1BOUND=rMAXABS(sSUM(sCONST(rMAX3(rT1,rT2,rT3))
     *           ,sCONST(rL1NORM(vT1))))
      ENDIF
      RETURN
      END
```

If $f^u(z) = \sum_{n=0}^{\infty} a_n z^n$, then this subroutine looks for the first coefficient different from zero, say a_{n_0}, and it follows that

$$f^u(z) = z^{n_0}(a_{n_0} + z^2 a_{n_0+2}) + a_{n_0+1} z^{n_0+1} + z^{n_0+3} h(z).$$ If a_{n_0}

and a_{n_0+2} have the same sign no improvement is possible (by these simple means) and we bound the function f^u by its ℓ_1-norm as indicated above. If the signs of a_{n_0} and a_{n_0+2} are opposite then we have the bound

$$\sup_{z \in \delta_1} |f^u(z)| \leq \max\{|a_{n_0}|, |a_{n_0} + a_{n_0+2}|,$$

$$((a_{n_0} + a_{n_0+2}\cos 2\varphi_0)^2 + (a_{n_0+2}\sin 2\varphi_0)^2)^{1/2}\} +$$

$$+ |a_{n_o+1}z^{n_o+1} + z^{n_o+3}h(z)|_1 \; ,$$

where $\delta_1 = \{z \in \mathbb{C} \mid |z| < 1 \text{ and } (|argz| < \varphi_o \text{ or } |argz-\pi| < \varphi_o)\}$. This bound follows readily by bounding explicitly the polynomial $a_{n_o} + a_{n_o+2}z^2$ on $\{|z| < 1, |argz| < \varphi_o\} \cup \{|z| < 1, |argz-\pi| < \varphi_o\}$ and by noting that $|z^{n_o}(a_{n_o} + a_{n_o+2}z^2)| \le |a_{n_o} + a_{n_o+2}z^2|$ for $|z| < 1$. We shall use the notation $|f^u|_B$ for $rL1BOUND(f^u)$.

We now explain the geometric organization of domains in the z-plane in detail. See Fig. 5.7.2

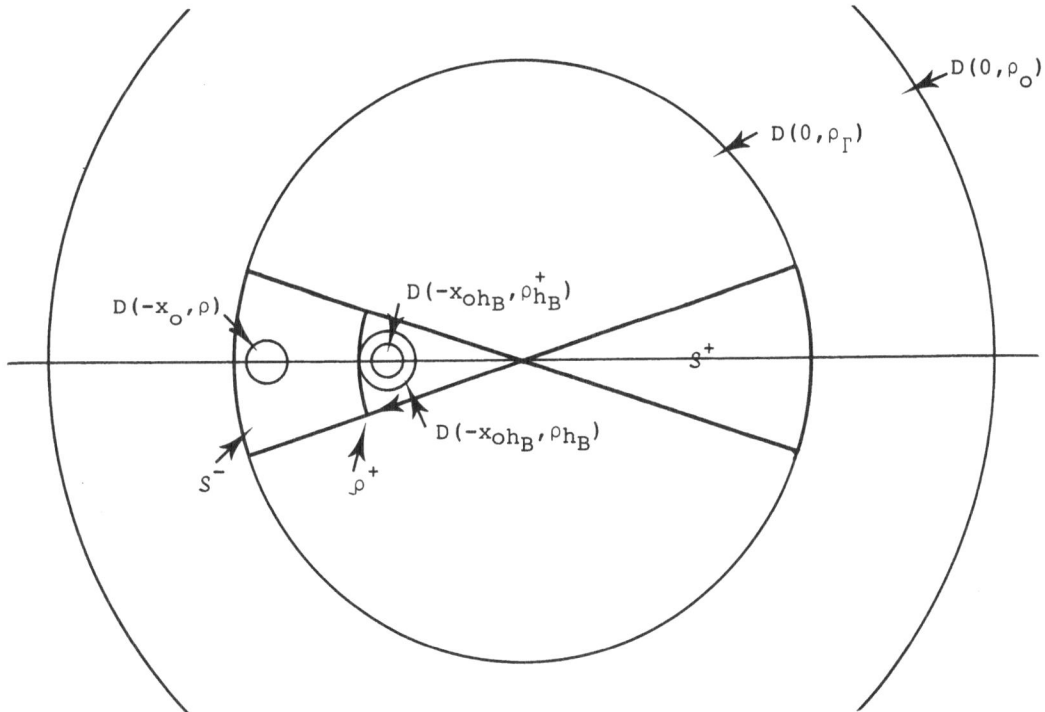

Fig. 5.7.2. Domains in the z-plane.

The constants $\varphi_0 \equiv$ sPHI$\emptyset$, $\rho_0 \equiv$ sRHOG$\emptyset$, $\rho_\Gamma \equiv$ sRHOGAMMA,

$x_{ohB} \equiv$ sX$\emptyset$hB, $\rho_{h_B} \equiv$ sRHOhB have already been defined in the main

program (see Section 5.5). Furthermore, $\rho_{h_B}^+ \equiv$ sRHOPhB with $\rho_{h_B}^+ = 0.008$

and $\rho^+ \equiv$ sRHOP with $\rho^+ = x_{ohB} + \rho_{h_B}^+$ (L47 and L55 of main program).

For x_0 and ρ we choose (L129, L130-132)

$$x_0 = -T_{-A}(-x_{oh}) ,$$

$$\rho = |T_{-A}(-x_{oh} + \rho_h z) - T_{-A}(-x_{oh})|_1 .$$

See L115-125 and Section 5.6 for T_{-A}. We have $\frac{1}{\rho_0}T_{-A}(-x_{oh}+\rho_h z) \equiv$

$\equiv$ v$\emptyset$ARG, $x_{oh} \equiv$ sXOh and $\rho_h \equiv$ sRHO h. ($x_{oh} = x_{oo} - \alpha_h$ and ρ_h are

also defined in the main program. See Sections 5.3 and 5.5.) The

disk $D(-x_0,\rho)$ contains the image of $D(-x_{oh},\rho_h)$ under the map

T_{-A} , for every A at hand. We have to check that $D(-x_0,\rho)$

and $D(-x_{ohB},\rho_{h_B}^+)$ are contained in the sector S^-. This we prove

in L133-140 and L144-151, where we show that $x_0 + \rho < \rho_\Gamma$,

$\sin\varphi_0 > \rho/x_0$ and $x_{ohB} + \rho_{h_B}^+ < \rho_\Gamma$, $\sin\varphi_0 < \rho_{h_B}^+/x_{ohB}$.

<u>Lemma 5.7.2.</u> (L153-212) <u>For every</u> $\{A,G_0\} \varepsilon D_K$ <u>and</u> $S^\pm$ <u>as</u>

<u>defined in</u> (5.2.5) <u>we have the inclusions</u> $G_0(S^\pm) \subset S^\mp$.

This estimate is quite delicate (though trivial) and needs a lot

of calculations. (Note that from $G_0(S^\pm) \subset S^\mp$ follows that

$\Gamma(S^\pm) \subset S^\pm$, and therefore $g((S)^2) \subset (S)^2$ as required.) We write

$$G_0(z) = -z + \frac{1}{\kappa}z^3 + z^3 r(z) ,$$

where $r(z) = 0(z)$. We need to bound $r(z)$ on the sectors $S^\pm$.

Unfortunately the bound which we get by using the subroutine

rL1BOUND is still not good enough and we therefore proceed as

shown in Fig. 5.7.3,

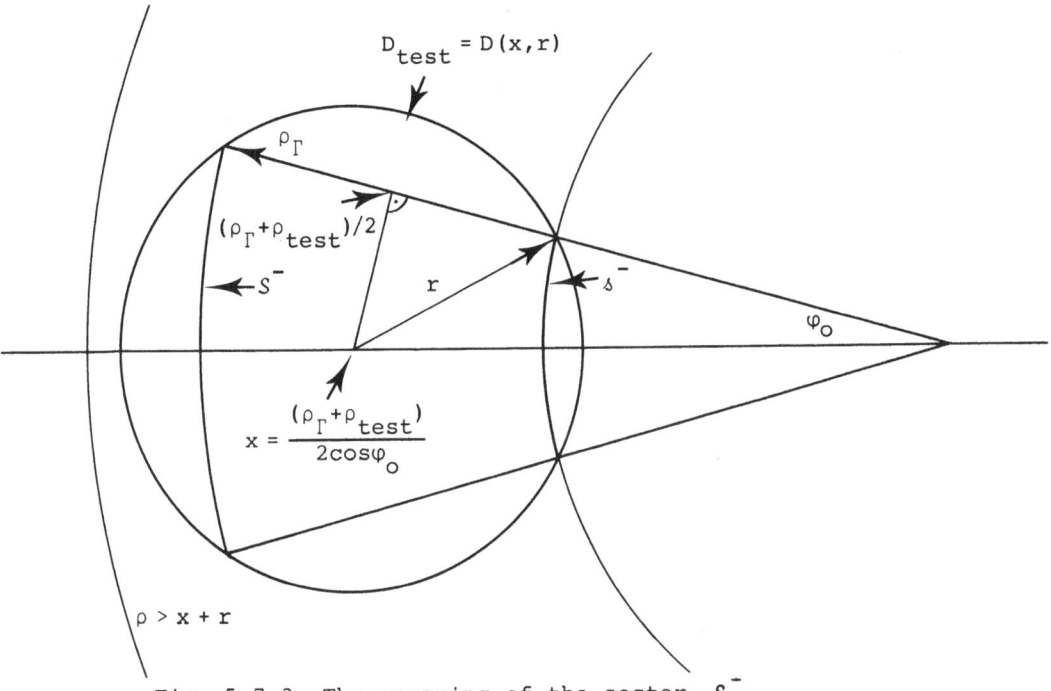

$$D_{test} = D(x,r)$$

$$\rho_\Gamma$$

$$(\rho_\Gamma + \rho_{test})/2$$

$$\leftarrow S^-$$

$$r$$

$$\phi_0$$

$$x = \frac{(\rho_\Gamma + \rho_{test})}{2\cos\phi_0}$$

$$\rho > x + r$$

Fig. 5.7.3. The covering of the sector S^-.

i.e. we cover the sectors $S^\pm$ by subsectors $s^\pm$ and disks $\pm D_{test}$ and establish bounds on the subsectors and the disks separately. We have (L176)

$$\sup_{z \in S^\pm} |r(z)| \le \max\{ \sup_{z \in s^\pm} |r(z)| , \sup_{z \in \pm D_{test}} |r(z)| \} ,$$

where $s^\pm = \{z \in S^\pm \,|\, |z| < \rho_{test}\}$ and $\pm D_{test} = D(\pm x, r)$, where

$$x = \frac{\rho_\Gamma + \rho_{test}}{2\cos\phi_0} \equiv sX ,$$

$$r = (x^2 + \rho_{test}^2 - 2x \cdot \rho_{test} \cdot \cos\phi_0)^{1/2} \equiv sR ,$$

(L155, L161-162). It is an easy calculation to show that

$$\sup_{z \in \delta^{\pm}} |r(z)| \le (\frac{1}{\rho_{test}})^3 \cdot \rho_o \cdot |G_o^u(\frac{\rho_{test}}{\rho_o}z) +$$

$$+ \frac{\rho_{test}}{\rho_o}z + \frac{1}{\kappa} \frac{\rho_{test}^3}{\rho_o} z^3|_B$$

and

$$\sup_{z \in \pm D_{test}} |r(z)| \le (\frac{1}{\rho_o})^2 |[\frac{1}{z^3}(G_o^u(z) + z + \frac{1}{\kappa} \rho_o^2 z)]$$

$$\circ (\frac{\pm x}{\rho_o} + \frac{r}{\rho_o}z)|_1$$

(L156-160 and L164-175). We now show that if $z = \rho e^{i\varphi} \epsilon S^{\pm}$ then $|G_o(z)| \le \rho$, i.e. lengths are contracted by applying G_o. We have

$$|G_o(z)| \le \rho \cdot |1 - \frac{1}{\kappa} \rho^2 \cos 2\varphi - \frac{i}{\kappa} \rho^2 \sin 2\varphi + z^2 r(z)|$$

$$\le \rho\{|1 - \frac{1}{\kappa} \rho^2 \cos 2\varphi - \frac{i}{\kappa} \rho^2 \cdot \sin 2\varphi| + \rho^2 |r(z)|\} .$$

We now check that

$$\frac{1}{\kappa} \rho_\Gamma^2 < 1 ,$$

(L178-179) and it follows that

$$|G_o(z)| \le \rho\{(1 - \frac{1}{\kappa} \rho^2 \cos 2\varphi) + \frac{\rho^2}{\kappa}\sin 2\varphi + \rho^2 |r(z)|\} .$$

We need to show that $-\frac{1}{\kappa} \rho^2 \cos 2\varphi + \frac{\rho^2}{\kappa}\sin 2\varphi + \rho^2 |r(z)| < 0$, which follows from $-\frac{1}{\kappa}\cos 2\varphi + \frac{1}{\kappa}\sin 2\varphi + |r(z)| < 0$. Note that for

$\varphi < \varphi_o$ the function $-\cos2\varphi + \sin2\varphi$ is increasing and therefore it suffices to show that

$$-\frac{1}{\kappa}(\cos2\varphi_o - \sin2\varphi_o) + \sup_{z\in S^{\pm}} |r(z)| < 0$$

(L178-183). This proves that G_o contracts lengths for $z\in S^{\pm}$. We now show that G_o contracts also angles. Let $z = x + iy\in S^{\pm}$ and denote $\hat{r}(z) = z^3 \cdot r(z)$. We have

$$y' := \text{ImG}_o(z) = -y + \frac{1}{\kappa}(3x^2y - y^3) + \text{Im}(\hat{r}(z)) ,$$

$$x' := \text{ReG}_o(z) = -x + \frac{1}{\kappa}(x^3 - 3xy^2) + \text{Re}(\hat{r}(z)) ,$$

and we want to show that $|y'/x'| < |y/x|$, if $|y/x| < \text{tg}\varphi_o$. We have

$$\left|\frac{y'}{x'}\right| = \left|\frac{1-\frac{1}{\kappa}(3x^2-y^2)-\frac{1}{y}\text{Im}(\hat{r}(z))}{x(1-\frac{1}{\kappa}(x^2-3y^2))-\text{Re}(\hat{r}(z))}\right| \cdot |y| .$$

We have already seen that $\rho_\Gamma^2/\kappa < 1$ and therefore it follows that $1 - \frac{1}{\kappa}(x^2 - 3y^2) > 0$. We next check that even

$$3\frac{\rho_\Gamma^2}{\kappa} < 1$$

(L185-186), from which it follows that $1 - \frac{1}{\kappa}(3x^2 - y^2) > 0$. Next we note that

$$|\text{Re}(\hat{r}(z))| \le |\hat{r}(z)| = |z|^3 \cdot |r(z)| = |x|^3|1 + i\frac{y}{x}|^3 \cdot |r(z)|$$

$$\leq |x|^3 (1 + tg^2\varphi_o)^{3/2} \cdot |r(z)| \ .$$

From this we get that

$$1 - \frac{1}{\kappa}(x^2 - 3y^2) - |x|^2(1 + tg^2\varphi_o)^{3/2} \cdot |r(z)| > 0 \ ,$$

provided

$$1 - \frac{\rho_\Gamma^2}{\kappa} - \rho_\Gamma^2(1 + tg^2\varphi_o)^{3/2} \cdot \sup_{z \in S^\pm} |r(z)| > 0 \ ,$$

which we check in L187-192. We have therefore

$$\left|\frac{y'}{x'}\right| \leq \frac{1 - \frac{1}{\kappa}x^2(3 - (\frac{y}{x})^2) + \frac{1}{|y|}|Im(\hat{r}(z))|}{1 - \frac{1}{\kappa}x^2(1 - 3(\frac{y}{x})^2) - x^2(1 + tg^2\varphi_o)^{3/2}|r(z)|} \cdot \left|\frac{y}{x}\right|$$

$$\leq \frac{1 - \frac{1}{\kappa}x^2 \cdot [3 - tg^2\varphi_o - \kappa\frac{1}{|x|^2|y|}|Im(\hat{r}(z))|]}{1 - \frac{1}{\kappa}x^2 \cdot [1 + \kappa(1 + tg^2\varphi_o)^{3/2} \cdot |r(z)|]} \cdot \left|\frac{y}{x}\right| \ .$$

For $Im(\hat{r}(z))$ we have

$$Im(\hat{r}(z)) = \frac{1}{2i}(\hat{r}(z) - \overline{\hat{r}(z)}) = \frac{1}{2i}(\hat{r}(z) - \hat{r}(\bar{z}))$$

$$= \frac{1}{2i} \int_{\bar{z}}^{z} \hat{r}'(\zeta) d\zeta = \frac{1}{2} \int_{-y}^{y} \hat{r}'(x + i\eta) d\eta \ ,$$

and therefore

$$|Im(\hat{r}(z))| \leq |y| \cdot \sup_{|\eta| < |y|} |\hat{r}'(x + i\eta)| \ .$$

We have $\hat{r}(z) = z^3 r(z)$ and therefore $\hat{r}'(z) = 3z^2 r(z) + z^3 r'(z)$.

We define $\hat{\hat{r}}(z) = 3r(z) + r'(z)$ and get

$$|\hat{r}'(x + i\eta)| = |x + i\eta|^2 \cdot |\hat{\hat{r}}(x + i\eta)| \leq |x|^2 (1 + tg^2\varphi_o)|\hat{\hat{r}}(z)|$$

and therefore $|Im(\hat{r}(z))| \leq |y||x|^2(1 + tg^2\varphi_o)|\hat{\hat{r}}(z)|$. Using this

we see that we have contraction of angles if

$$\frac{1 - \frac{1}{\kappa}x^2(3 - tg^2\varphi_o - \kappa(1 + tg^2\varphi_o)\,\sup_{z\in S^\pm}|\hat{\hat{r}}(z)|)}{1 - \frac{1}{\kappa}x^2(1 + \kappa(1 + tg^2\varphi_o)^{3/2} \cdot \sup_{z\in S^\pm}|r(z)|)} < 1 \quad,$$

i.e. if

$$tg^2\varphi_o + \kappa(1 + tg^2\varphi_o)^{3/2} \sup_{z\in S^\pm}|r(z)| + \kappa(1 + tg^2\varphi_o) \sup_{z\in S^\pm}|\hat{\hat{r}}(z)| < 2$$

(check in L208-212). The only lines which we need to explain in

order to understand the invariance of the sectors $S^\pm$ are L193-

211. In these lines we establish the bound on $\sup_{z\in S^\pm}|\hat{\hat{r}}(z)|$. We

have

$$\sup_{z\in S^\pm}|\hat{\hat{r}}(z)| \leq 3 \cdot \sup_{z\in S^\pm}|r(z)| + \sup_{z\in S^\pm}|z\cdot r'(z)|$$

(L207) and we still need to bound $\sup_{z\in S^\pm}|z\cdot r'(z)|$. This is done in

the same spirit as in the calculation for $\sup_{z\in S^\pm}|r(z)|$ above. We

have

$$\sup_{z\in S^\pm}|z\cdot r'(z)| \leq \max\{\sup_{z\in\delta^\pm}|z\cdot r'(z)|, \quad \sup_{z\in{}^\pm D_{rest}}|z\cdot r'(z)|\} \quad,$$

where

$$\sup_{z \in S^\pm} |z \cdot r'(z)| \leq \rho_{rest} \cdot |[\frac{1}{z^3}(G_o^u(z) + z - \frac{1}{\kappa} \rho_o^2 z^3)]' \circ (\frac{\rho_{rest}}{\rho_o} z)|_B$$

and

$$\sup_{z \in \pm D_{rest}} |z \cdot r'(z)| \leq |\sqrt{\frac{\rho_\Gamma}{\rho_o}} \{z \cdot ([\frac{1}{z^3}(G_o^u(z) + z - \frac{1}{\kappa} \rho_o^2 z^3)]' \circ$$

$$\circ (\sqrt{\frac{\rho_\Gamma}{\rho_o}} z)) \} \circ \{\frac{\pm x}{\sqrt{\rho_o \rho_\Gamma}} + \frac{r}{\sqrt{\rho_o \rho_\Gamma}} \cdot z\}|_1 \quad ,$$

with x and r as above (L193-194 and L195-205). This completes the estimates which are necessary to prove that
$G_o(S^\pm) \subset S^\mp$. We next verify that the sectors $S^\pm$ are attracted
by z = 0 (L216-222). We have shown in Section 2 that this fol-
lows from $Re((\frac{1}{\Gamma(z)})^2 - \frac{1}{z^2}) \geq L_5 > 0$. We set $\Gamma(z) = z \cdot H(z)$,
where $\Gamma^u(z) \equiv GAMMA$, $H^u(z) \equiv vETA$. Define now

$$R(z) = (\frac{1}{\Gamma(z)})^2 - \frac{1}{z^2} . \tag{5.7.4}$$

We have that $R(z) = \frac{4}{\kappa} + O(z)$, and therefore $R(z)$ is analytic
near z = 0, and $R(0) = \frac{4}{\kappa} > 0$. In fact, since we calculate
$H^u(z) \equiv vETA$ and $(H^u(z))^{-2} \equiv vI2ETA$ (L61-62 and L66-69) without error termination it follows that $R(z)$ is analytic on
$D(0, \rho_\Gamma)$. An easy calculation shows now that

$$\inf_{z \in S^\pm} Re\ R(z) \geq R(0) - \sup_{z \in S^\pm} |R(z) - R(0)|$$

$$\geq R(0) - \sup_{z \in D(0, \rho_\Gamma)} |R(z) - R(0)|$$

$$= R(0) - \sup_{\varphi} |R(\rho_\Gamma e^{i\varphi}) - R(0)|$$

$$\geq R(0) - (\frac{1}{\rho_\Gamma})^2 \cdot |(\frac{1}{H^u(z)})^2 - 1 - R(0)\rho_\Gamma^2 z^2|_1 \ .$$

For sake of completeness we also explain the bounds given in L226-250. We define the sector δ by

$$\delta = \{z \in S^- | |z| < \rho^+\} \ . \tag{5.7.5}$$

Then

$$\inf_{z \in \delta} |\text{Re}(R(z))| \geq R(0) - (\frac{1}{\rho^+})^2 |(\frac{1}{H^u(\rho^+ z/\rho_\Gamma)})^2 - 1 - R(0)(\rho^+)^2 z^2|_1$$

$$\equiv sL5 \tag{5.7.6}$$

(L226-232), where $\rho^+ \equiv$ sRHOP as in Fig. 5.7.2. We use sRHO1 $\equiv \rho^+/\rho_\Gamma$ throughout the program. (sRHO1 has already been defined in the main program. See Section 5.5.) sL6 $\equiv \cos2\varphi_0$ and sL7 $\equiv \sin2\varphi_0$ (L234 and L235) are two constants which we shall need below. Furthermore we shall need

$$\frac{1}{r} = \{\cos2\varphi_0 (1 - |(\frac{1}{H^u(\rho^+ z/\rho_\Gamma)})^2 - 1|_B)$$

$$- \sin2\varphi_0 \cdot |(\frac{1}{H^u(\rho^+ z/\rho_\Gamma)})^2 - 1|_B\}^< \ , \tag{5.7.7}$$

where $\{---\}^<$ means that we need a lower bound of the expression in the parenthesis. We have $sZ \equiv r$ (L237-241). In L243-244 we calculate

$$\text{sETA} \equiv (|(\frac{1}{H^u(\rho^+ z/\rho_\Gamma)})^2|)^{-1/2} , \qquad (5.7.8)$$

This completes our discussion of domain questions.

5.7.2 Solving Ecalle's equation (general outline)

We explain now how on these domains (5.3.2) is solved. The corresponding lines are L47-57, L71-90, L246-249 and L252-301, and the corresponding calculation are carried through in the sub-routines CREDUCE (which in turn calls FREDUCE), CxFØ, PREPBOREL, FSUMBOREL, FADDGENERAL (which in turn calls FBNDIM and FBNDPREIM) and FCOMPLETE. We explain in the following our strategy for solving (5.3.2) in detail and explain also all the subroutines which are involved in the calculation. Note that L47-57, L71-90, L246-249 and L252-301 are the only lines of FvKI $\equiv$ K which remain to be explained, so that after the discussion of these lines the bound $\epsilon = \|K(\{A,G\}) - \{A,G\}\|$ will be defined.

Notation : Let f be a function analytic on some disk centered at z = 0 and vf the vector representing the function f on the computer. We have seen that f has a natural decomposition into a polynomial part f_p a higher order part Δf_H and a general part Δf_G which are represented on the computer by the polynomial part of vf, the bound rH(vf) and the bound rG(vf). We shall extend the notation in the following to functions f which are analytic on the sector δ (see (5.7.5)). We shall write for such a function f,

$$f = f_P + \Delta f_G + \Delta f_H \ ,$$

meaning that f can be decomposed in a polynomial part f_P and

functions Δf_G , Δf_H which are analytic on δ and satisfy

$$\Delta f_G(z) = z^{n(G)} f_G(z) \ ,$$

$$\Delta f_H(z) = z^{n(H)} f_H(z) \ ,$$

where $n(G), n(H) \in \mathbb{N}$ may depend on f and will be explicitly

given below. Furthermore, one should think of f_G and f_H as being

"small", i.e. one should think of $\sup_{z \in \delta} |f_G(z)|$ or $\sup_{z \in \delta} |f_H(z)|$ as

being small compared to $\sup_{z \in \delta} |f_P(z)|$. We shall also use the notation

$f_{P,thin}$ to express that the polynomial part of f is represented

on the computer by a polynomial, having "scalar" coefficients

which contain only one "rep", i.e. an interval of length zero.

We describe now the practical aspects of solving Ecalle's

equation with error bounds. (These estimates in fact occupy a

suitable fraction of the program, since the standard arithmetic

package does not provide such specialized bounds.) We consider

first the equation

$$F_o(\Gamma(z)) - F_o(z) = R(z) = -2 \ , \qquad\qquad (5.7.9)$$

where $\Gamma \equiv$ vGAMMA is given and where we set vRHS $= R(z)$ in

L257-258. Furthermore, in the sense defined above, we decompose
$\Gamma = \Gamma_P + \Delta\Gamma_G + \Delta\Gamma_H$, $\Delta\Gamma_G = O(z^3)$, $\Delta\Gamma_H = O(z^{N+1})$, where N is the polynomial degree which has been choosen. (N = 67 in the case where IFLAG = 0, i.e. where FvKI $\equiv$ K.) Furthermore,

$$\sup_{z \in \delta} \left| \frac{\Delta\Gamma_G}{(z/\rho^+)^3} \right| \leq \rho_\Gamma (\rho^+/\rho_\Gamma)^3 |\Delta\Gamma_G^u|_1 \equiv \text{SGGAMMA} , \qquad (5.7.10)$$

$$\sup_{z \in \delta} \left| \frac{\Delta\Gamma_H}{(z/\rho^+)^{N+1}} \right| \leq \rho_\Gamma (\rho^+/\rho_\Gamma)^{N+1} |\Delta\Gamma_H^u|_1 \equiv \text{SHGAMMA} , \qquad (5.7.11)$$

where we have used the notation $\Gamma^u = \Gamma_P^u + \Delta\Gamma_G^u + \Delta\Gamma_H^u$ in the sense defined above and where therefore on the computer

$$|\Delta\Gamma_G^u|_1 = \text{rG(vGAMMA)} ,$$

$$|\Delta\Gamma_H^u|_1 = \text{rH(vGAMMA)} .$$

SGGAMMA and SHGAMMA are calculated in L246-247 and L248-249 respectively.

We give now in form of a table the logical relation of the subroutines mentioned above, together with a glossary of their action.

Table 5.7.1

Ansatz : $F_o = F_{sing} + F_{reg}$, where

$$F_{sing} = a_o/z^2 + c_o \log z^2 + d_o z$$

Equations : $F_{reg}(\Gamma) - F_{reg} = J \quad (= \mathcal{O}(z^4))$

$$J = R - (F_{sing}(\Gamma) - F_{sing})$$

$$= J_{P,thin} + \Delta J_G + \Delta J_H$$

$\left.\begin{array}{c} \\ \\ \\ \\ \\ \end{array}\right\}$ FREDUCE

Ansatz : $F_{reg} = F + \Delta F$

Equations : $F(\Gamma_P) - F = J_{P,thin}$

$$\Delta F(\Gamma) - \Delta F = \Delta J_G + \Delta J_H - (F(\Gamma) - F(\Gamma_P))$$

Solutions : $\Gamma_P, J_{P,thin} \rightarrow \hat{\Gamma}_P(x), \hat{J}_{P,thin}(x)$ (formal power

series) and bound :

$$\left| ((F_B)_{normal})_{Borel}(t) \right| \le C_{F\emptyset} \exp(E_{F\emptyset}|t|)$$
$$form transform$$

$\left.\begin{array}{c} \\ \\ \\ \\ \\ \\ \\ \\ \\ \end{array}\right\}$ CREDUCE

Solutions : i) $\hat{F}(\hat{\Gamma}_P(x)) - \hat{F}(x) = \hat{J}_{P,thin}(x)$ (formal power
    series) $\}$ CxFØ

  ii) $\hat{F}(x)$, $C_{F\emptyset}$ and $E_{F\emptyset}$ (i.e. $C_{\mathcal{L}}, R_{\mathcal{L}}) \rightarrow h_B^u$ $\left.\begin{array}{c} \\ \\ \end{array}\right\}$ FPREPBOREL
    +
    FSUMBOREL

 iii) $\Delta J_{F\emptyset,G} + \Delta J_{F\emptyset,H} = -(F(\Gamma) - F(\Gamma_P))$ FBNDIM

 iv) $\Delta J_{tot,G} = \Delta J_G + \Delta J_{F\emptyset,G}$

    $\Delta J_{tot,H} = \Delta J_H + \Delta J_{F\emptyset,H}$

  v) $\Delta F(\Gamma) - \Delta F = \Delta J_{tot,G} + \Delta J_{tot,H}$ $\left.\begin{array}{c} \\ \\ \end{array}\right\}$ FBNDPREIM
    where $\Delta F = \Delta F_G + \Delta F_H$

$\left.\begin{array}{c} \\ \\ \\ \\ \\ \\ \end{array}\right\}$ FADDGENERAL

  [vi) $F_o = F_{sing} + F_{reg}$ $\}$ FCOMPLETE]

The idea is the following. Since we are working on the computer
with vectors, we always have to treat "polynomial parts" and
"parts for which we only know bounds". The Borel summation procedure works perfectly well on "polynomial parts" since the method
takes – through resummation – into account the cancellations occurring between the coefficients of the formal power series solution.
However, on the "parts for which we only know bounds" such cancellations can not be taken into account and we use directly the
Laplace-representation (2.1.3) of the solution to treat those
parts. We therefore observe the splitting in "power series part"
and "bounds" throughout our construction of $vhB \equiv h_B^u$. More pre-
cisely, we proceed as described in Table 5.7.1.

i) We split off (using our general subroutine package of Section 4) the part F_{sing} of the solution (i.e. we determine
   a_o, c_o, d_o and J). This is done in the subroutine FREDUCE.
   It furthermore splits J into a polynomial part $J_{P,thin}$
   and bounds ΔJ_G, ΔJ_H.

ii) We construct the solution F_{reg} as the sum of two parts, F
    and ΔF. It is the function F which we construct by Borel
    summation using Loeffel's method. For ΔF we only establish
    bounds using the representation (2.1.3) for solutions of equa-
    tions of Ecalle's type. More precisely we proceed as follows.
    We first construct F which solves $F(\Gamma_p) - F = J_{P,thin}$.
    (Note that we take only the polynomial part Γ_p in the
    argument of F. Note furthermore that F is not a polynomial.)
    We then construct the right hand side of the equation which

must be satisfied by ΔF and bound ΔF. The subroutine CREDUCE renders the power series $\hat{J}_{P,thin}(x)$ corresponding to the function $J_{P,thin}$, furthermore the bounds on ΔJ_G and ΔJ_H and the bounds on the Borel transform f_B of $f = F((\frac{\kappa}{4}z)^{1/2})$. (L259-261 of FvKI.)

iii) Given the formal power seires $\hat{\Gamma}_P(x)$ and $\hat{J}_{P,thin}(x)$ the subroutine CxFØ calculates the formal power series solution $\hat{F}(x)$ of Ecalle's equation (at least a finite number of coefficients to it). (L268 of FvKI.)

iv) Using the formal power series $\hat{F}(x)$ and the bounds C_f and E_f of f_B (respectively C_ℓ and R_ℓ of $f_{B,\ell}$) as an input, the subroutine FSUMBOREL (together with FPREPBOREL and some additional functions) calculates the contribution of the function F to $vhB \equiv h_B^u$. (L277, L292 of FvKI.)

v) Using the bounds ΔJ_G and ΔJ_H and furthermore C_f and E_f as an input, the subroutine FBNDIM calculates the contribution to the right hand side of the equation for ΔF which comes from the fact that we have used Γ_P in the argument of F instead of Γ. This is done as a substep in FADDGENERAL, where afterwards, using FBNDPREIM, ΔF is estimated. This leads to the addition of a bound on a general term in $vhB \equiv h_B^u$. (L296-297 of FvKI.)

vi) The construction of vhB is completed by adding the contribution of the part F_{sing} to $vhB \equiv h_B^u$. This is done in the subroutine FCOMPLETE.   (L301 of FvKI.)

This completes the general exposition of the subroutine

package which carries through the solution of Ecalle's equation

and its Borel summation.

5.7.3 Bounds and power series solution of Ecalle's equation

We proceed now to give the bounds in detail. We discuss first

the subroutine CREDUCE and FREDUCE and give again first the

listings and explain then their contents in detail.

```
 1:        SUBROUTINE CREDUCE(vRHS,vA0,vC0,vD0,sA0,sC0,sD0,vxJ,xJ,sGJ,sHJ
 2:     *                    ,IFLAG,sJ00,sJ01,sJ02,sGDZJ,sHDZJ
 3:     *                    ,vI2ETA,sCF0,sEF0)
 4:       INCLUDE BIGN.STUFF
 5:       COMMON/ECALLE/sC3
 6:       COMMON/DOMAIN/sRHOG0,sRHOGAMMA,sXOh,sCENTh,sRHOh,sCENTG,sRHOG
 7:     *             ,sRHOhB,sRHOPhB,sXOhB,sPHIO,sxKAPPA,sKAPPA
 8:     *             ,sRHOP,sRHO1
 9:       COMMON/SWITCH/ITEST
10:       DIMENSION vRHS(1),vA0(1),vC0(1),vD0(1),vxJ(1),xJ(1),vI2ETA(1)
11:       DIMENSION vT1(NDEG)
12:
13:
14:C  Reduction to regular inhomogeneous part
15:C  =======================================
16:
17:       CALL FREDUCE(vRHS,vA0,vC0,vD0,sA0,sC0,sD0,vxJ,sGJ,sHJ
18:     *             ,IFLAG,sJ00,sJ01,sJ02,sGDZJ,sHDZJ)
19:
20:
21:C  Produce unnormalized power series
22:C  =================================
23:
24:       CALL CvTOxe(vxJ,xJ,eT1)
25:       ITEST=0
26:       CALL FsDILAT(xJ,sINV(sRHOGAMMA),xJ)
27:       ITEST=1
28:
```

```
29:
30:C    Calculate bounds of Borel function
31:C    ==================================
32:
33:C    Calculate bounds for vxJH
34:C    -------------------------
35:       sEJH=sUPPER(sQUOT(sxKAPPA,sPROD(sPROD(sFOUR,sRHOGAMMA)
36:      *      ,sRHOGAMMA)))
37:       rT1=rL1NORM(vxJ)
38:       sCJH=sUPPER(sPROD(sPROD(sEJH,sEJH),sCONST(rT1)))
39:
40:C    Compute bounds on vI2ETA
41:C    ------------------------
42:       sEI2ETA=sEJH
43:       CALL FEQUAL(vI2ETA,vT1)
44:       vT1(1)=sZERO
45:       vT1(2)=sZERO
46:       vT1(3)=sZERO
47:       vT1(4)=sZERO
48:       rT1=rL1NORM(vT1)
49:       sCI2ETA=sUPPER(sPROD(sPROD(sEI2ETA,sEI2ETA),sCONST(rT1)))
50:
51:C    Compute alternate bounds for vxJH
52:C    ---------------------------------
53:       sC31=sSUM(sONE,sC3)
54:       sEJHT=sLOWER(sDIFF(sSUM(sEJH,sPROD(sCI2ETA,sC31)),sONE))
55:       sRHO=sSQRT(sQUOT(sxKAPPA,sPROD(sFOUR,sEJHT)))
56:       CALL FsDILAT(vxJ,sQUOT(sRHO,sRHOGAMMA),vT1)
57:       rT1=rL1NORM(vT1)
58:       sCJHT=sUPPER(sPROD(sPROD(sEJHT,sEJHT),sCONST(rT1)))
59:
60:C    Compute bounds for vF0
61:C    ----------------------
62:       sC31=sSUM(sONE,sC3)
63:       sT1=sPROD(sCI2ETA,sC31)
64:       IF(rMAXABS(sT1).LT.rONE)WRITE(6,*)´error in CREDUCE
65:C..compute sEF0
66:       sEF0=sUPPER(sSUM(sEJH,sT1))
67:C..compute sCF0
68:       sT2=sPROD(sC3,sCJHT)
69:       sT3=sQUOT(sPROD(sC3,sCJH),sPROD(sT1,sC31))
70:       sT4=sSUM(sONE,sPROD(sC3,sSUM(sONE,sT1)))
71:       sCF0=sUPPER(sSUM(sT2,sPROD(sT3,sT4)))
72:
73:
74:       RETURN
75:       END
```

FREDUCE

```
  1:          SUBROUTINE FREDUCE(vRHS,vA0,vC0,vD0,sA0,sC0,sD0,vxJ,sGJ,sHJ
  2:        *                  ,IFLAG,sJ00,sJ01,sJ02,sGDZJ,sHDZJ)
  3:          INCLUDE BIGN.STUFF
  4:          COMMON/DOMAIN/sRHOG0,sRHOGAMMA,sXOh,sCENTh,sRHOh,sCENTG,sRHOG
  5:        *             ,sRHOhB,sRHOPhB,sXOhB,sPHIO,sxKAPPA,sKAPPA
  6:        *             ,sRHOP,sRHO1
  7:          COMMON/REDUCE/sG03,sG04,sG05,sGA3,sGA33,sGA5,sGA6
  8:          DIMENSION vRHS(1),vA0(1),vC0(1),vD0(1),vxJ(1)
  9:          DIMENSION vJ(NDEG),vDZJ(NDEG),vxDZJ(NDEG)
 10:
 11:
 12:C   Calculation of irregular inhomogeneous part
 13:C   ============================================
 14:
 15:          sA0=sNEG(sQUOT(sPROD(sPOWER(sRHOG0,2),sHALF),sG03))
 16:          sC0=sNEG(sSUM(sQUOT(sTHREE,sFOUR),sPROD(sHALF,sQUOT(sG05
 17:        *    ,sPOWER(sG03,2)))))
 18:          sD0=sNEG(sQUOT(sPROD(sHALF,sG04),sPROD(sRHOG0,sG03)))
 19:
 20:
 21:C   Calculate regular inhomogeneous part  (and its derivative)
 22:C   ==========================================================
 23:
 24:C   Store vRHS
 25:C   ----------
 26:          CALL FEQUAL(vRHS,vJ)
 27:
 28:C   Add terms with sA0
 29:C   ------------------
 30:          CALL FsMULADD(vA0,sNEG(sA0),vJ)
 31:
 32:C   Add term with sC0
 33:C   -----------------
 34:          CALL FsMULADD(vC0,sNEG(sPROD(sTWO,sC0)),vJ)
 35:
 36:C   Add term with sD0
 37:C   -----------------
 38:          CALL FsMULADD(vD0,sNEG(sD0),vJ)
 39:
 40:C   Normalize properly
 41:C   ------------------
 42:          vJ(1)=sZERO
 43:          vJ(2)=sZERO
 44:          vJ(3)=sZERO
 45:          vJ(4)=sZERO
 46:
 47:C   If IFLAG=1 calculate derivative
 48:C   -------------------------------
 49:          IF(IFLAG.EQ.1)THEN
 50:            sTG=sQUOT(sCONST(rG(vJ)),sRHOGAMMA)
 51:            rRHO1=rMAXABS(sRHO1)
 52:            rTG3=rMAXABS(sPROD(sCONST(rSUP(rRHO1,4,1)),sTG))
 53:            rTG4=rMAXABS(sPROD(sCONST(rSUP(rRHO1,5,1)),sTG))
 54:            rTG5=rMAXABS(sPROD(sCONST(rSUP(rRHO1,6,1)),sTG))
 55:            sTG6=sPROD(sCONST(rSUP(rRHO1,7,1)),sTG)
 56:            CALL FDZsDILAT(vJ,sRHO1,vDZJ)
 57:            vDZJ(1)=sZERO
 58:            CALL FsMULT(vDZJ,sINV(sRHOGAMMA),vDZJ)
 59:            vDZJ(NERR)=error(rTG3,rH(vDZJ))
 60:          ENDIF
 61:
 62:
 63:C   Split off central part; calculate sGJ and sHJ
 64:C   =============================================
 65:
```

FREDUCE

```
 66:C  Choose central values of coefficients and calculate difference
 67:C  ---------------------------------------------------------------
 68:       CALL FSETZERO(vxJ)
 69:       DO 1 I=1,N1
 70:       vxJ(I)=saCONST(aCENT(vJ(I)))
 71:       vJ(I)=sDIFF(vJ(I),vxJ(I))
 72:     1 CONTINUE
 73:
 74:C  If IFLAG=1 calculate derivative
 75:C  -------------------------------
 76:       IF(IFLAG.EQ.1)THEN
 77:         CALL FDZsDILAT(vxJ,sRHO1,vxDZJ)
 78:         CALL FsMULT(vxDZJ,sINV(sRHOGAMMA),vxDZJ)
 79:       ENDIF
 80:
 81:C  Scale down
 82:C  ----------
 83:       CALL FsDILAT(vJ,sRHO1,vJ)
 84:
 85:C  Collect higher order bounds in sHJ
 86:C  ----------------------------------
 87:       sTH=sCONST(rH(vJ))
 88:       sT1=sABS(vJ(N1-2))
 89:       sT2=sABS(vJ(N1-1))
 90:       sT3=sABS(vJ(N1  ))
 91:       sHJ=sUPPER(sSUM(sSUM(sSUM(sT1,sT2),sT3),sTH))
 92:
 93:C  Collect bounds on difference in sGJ
 94:C  -----------------------------------
 95:       sTG=sCONST(rG(vJ))
 96:       vJ(N1-2)=sZERO
 97:       vJ(N1-1)=sZERO
 98:       vJ(N1  )=sZERO
 99:       vJ(NERR)=sZERO
100:       sGJ=sUPPER(sSUM(sCONST(rL1NORM(vJ)),sPROD(sTG,sPOWER(sRHO1,3))))
101:
102:
103:C  Return if IFLAG=0
104:C  -----------------
105:       IF(IFLAG.EQ.0)RETURN
106:
107:
108:C  Calculate sGDZJ and sHDZJ
109:C  =========================
110:
111:C  Calculate difference
112:C  --------------------
113:       CALL FMINUS(vxDZJ,vDZJ)
114:
115:C  Collect higher order bounds in sHDZJ
116:C  ------------------------------------
117:       sTH=sCONST(rH(vDZJ))
118:       sT1=sABS(vDZJ(N1-3))
119:       sT2=sABS(vDZJ(N1-2))
120:       sT3=sABS(vDZJ(N1-1))
121:       sT4=sABS(vDZJ(N1  ))
122:       sHDZJ=sUPPER(sSUM(sSUM(sSUM(sSUM(sT1,sT2),sT3),sT4),sTH))
123:
124:C  Collect bounds on difference in sJ00,sJ01,sJ02 and sGDZJ
125:C  -------------------------------------------------------
126:       sTG=sCONST(rG(vDZJ))
127:       vDZJ(N1-3)=sZERO
128:       vDZJ(N1-2)=sZERO
129:       vDZJ(N1-1)=sZERO
130:       vDZJ(N1  )=sZERO
```

FREDUCE

```
131:        vDZJ(NERR)=sZERO
132:        sJ00=sQUOT(sSUM(vDZJ(4),DCMPLX(-rTG3,rTG3)),sPOWER(sRHOP,3))
133:        sJ01=sQUOT(sSUM(vDZJ(5),DCMPLX(-rTG4,rTG4)),sPOWER(sRHOP,4))
134:        sJ02=sQUOT(sSUM(vDZJ(6),DCMPLX(-rTG5,rTG5)),sPOWER(sRHOP,5))
135:        vDZJ(4)=sZERO
136:        vDZJ(5)=sZERO
137:        vDZJ(6)=sZERO
138:        sGDZJ=sUPPER(sSUM(sCONST(rL1BOUND(vDZJ)),sTG6))
139:
140:
141:        RETURN
142:        END
```

Since the subroutine CREDUCE calls in a first step (L17-18) FREDUCE we discuss this subroutine first. Line numbers refer to lines in FREDUCE if not stated otherwise.

In L15-18 we split off the irregular part of the solution. If we write $G_o^u(z) = -z + g_3 z^3 + g_4 z^4 + g_5 z^5 + 0(z^6)$, then

$$a_o = -\rho_o^2/(2g_3) \equiv \text{sA}\varnothing$$

$$d_o = -\rho_4/(2\rho_o g_5) \equiv \text{sD}\varnothing$$

$$c_o = -\frac{3}{4} - \frac{1}{2} g_5/(g_3)^2 \equiv \text{sC}\varnothing \ ,$$

where $g_3 \equiv \text{sG}\varnothing 3$, $g_4 \equiv \text{sG}\varnothing 4$ and $g_5 \equiv \text{sG}\varnothing 5$ have already been calculated L49-52 of FvKI. (These are variables in the COMMON/REDUCE/.) The reason why we use the coefficients of G_o and not those of Γ is that a_o, d_o, c_o can be calculated with more precision that way. Next we need to calculate the function J, i.e. we want to bring (5.7.9) to the form

$$F_{reg}(\Gamma(z)) - F_{reg}(z) = J(z) \ , \qquad\qquad (5.7.12)$$

where

$$J(z) = R(z) - (F_{sing}(\Gamma(z)) - F_{sing}(z))$$

$$= R(z) - a_o \left(\left(\frac{1}{\Gamma(z)}\right)^2 - \frac{1}{z^2} \right) - 2c_o \log\left(\frac{\Gamma(z)}{z}\right)$$

$$- d_o \cdot (\Gamma(z) - z) \ . \qquad\qquad (5.7.13)$$

We have $J^u(z) = J(\rho_\Gamma z)$. We define now the functions

$$A_o(z) = (1/\Gamma(z))^2 - 1/z^2 ,$$

$$C_o(z) = \log(\Gamma(z)/z) = \log H(z) ,$$

$$D_o(z) = \Gamma(z) - z ,$$

$A_o^u(z) = A_o(\rho_\Gamma z)$, $C_o^u(z) = C_o(\rho_\Gamma z)$, $D_o^u(z) = D_o(\rho_\Gamma z)$. We have therefore

$$A_o^u(z) = (1/\rho_\Gamma)^2 \cdot [\frac{1}{z^2} ((1/H^u(z))^2 - 1)] \equiv vA\emptyset \qquad (5.7.14)$$

$$C_o^u(z) = \log H^u(z) \equiv vC\emptyset \qquad (5.7.15)$$

$$D_o^u(z) = \rho_\Gamma \cdot (\Gamma^u(z) - z) \equiv vD\emptyset \qquad (5.7.16)$$

and $J^u(z) = R^u(z) - a_o \cdot A_o^u(z) - 2 \cdot c_o \cdot C_o^u(z) - d_o \cdot D_o^u(z)$. The functions A_o^u, C_o^u and D_o^u have already been calculated in L73-80, L84 and L88-90 of FvKI. The function J^u is calculated in L26-45. In L42-45 we use the fact that we know that $J^u(z) = 0(z^4)$. Since IFLAG = 0 we skip for the moment L49-60 and pass to the discussion of L66-100. Here we split the function J in $J_{P,thin} + \Delta J_G + \Delta J_H$, where

$$\Delta J_G(z) = 0(z^4) ,$$

$$\qquad (5.7.17)$$

$$\Delta J_H(z) = 0(z^{N-3}) ,$$

with (L100, L91)

$$\sup_{z \in \delta} \left| \frac{\Delta J_G(z)}{(z/\rho^+)^4} \right| \leq |\Delta J_G^u(\rho^+ z/\rho_\Gamma)|_B \equiv sGJ \ , \qquad (5.7.18)$$

$$\sup_{z \in \delta} \left| \frac{\Delta J_H(z)}{(z/\rho^+)^{N-3}} \right| \leq |\Delta J_H^u(\rho^+ z/\rho_\Gamma)|_B \equiv sHJ \ , \qquad (5.7.19)$$

<u>Remark</u> : We choose $\Delta J_H(z) = O(z^{N-3})$ and not $O(z^{N+1})$ as the reader might have expected. The reason is the following : We want to split off a polynomial part $J_{P,thin}$, and not J_P only, in order to make the power series computation for the Borel summation part as precise as possible (see below). We have therefore to sum the difference between J_P and $J_{P,thin}$ (which is essentially the width of the scalars describing J_P) either into the bound ΔJ_G or into the bound ΔJ_H. The scalars corresponding to the orders $O(z^4)$ up to $O(z^{N-3})$ are only blown up during the calculation of the function J due to up and down rounding, whereas the orders $O(z^{N-2})$ up to $O(z^N)$ contain also contributions from bounds on higher order functions and have a considerably larger width. (See the subroutine FSHFTL in Section 4.) We want to keep as long as possible the information about the order of the bounds and collect therefore the orders $O(z^{N-2})$ up to $O(z^N)$ into a new "higher order bound" which is $O(z^{N-2})$. Our choice $\Delta J_H(z) = O(z^{N-3})$ is a compromise between "making the program efficient and the bounds optimal". (See below.) All this being said we shall henceforth not readdress this question.

In L66-72 we choose $J^u_{P,thin}$ and store the difference between this polynomial and the polynomial J^u_P. Since we need bounds only on δ we scale these differences down, L83, and collect them in L87-91 (higher orders) and in L95-100. The bound in L100 is scaled down by an additional power $(\rho_1)^3$, because our subroutine FsDILAT supposes the general parts to be of order $O(z)$ but we know in this case that it is $O(z^4)$. The remaining lines of FREDUCE are only used in the case IFLAG = 1 which will be explained below. This completes our discussion of the subroutine FREDUCE and we continue our discussion of CREDUCE. In L24-27 we transform $J^u_{P,thin} \equiv vxJ$ into a formal power series $xJ \equiv \hat{J}_{P,thin}(x)$ and in the remaining lines L35-71 we calculate the bounds C_f and E_f of the Borel transform of the solution. This is done as follows :

Let $j(z) = J((\frac{\kappa}{4}z)^{1/2})$ be the normal form of J. j is analytic on $D_2(0, \frac{4}{\kappa}\rho_\Gamma^2) \subset \mathbb{C}_2$, and $j(z) = O(z^2)$. We write $j(z) \equiv z \cdot \hat{j}(z)$. For the Borel transform $j_B(t)$ we get

$$j_B(t) = 1 * \hat{j}_B(t) \equiv \int_o^t \hat{j}_B(t_1) dt_1 ,$$

where $*$ means convolution. Suppose now that we are given the bound

$$|\hat{j}_B(t)| \leq C_{\hat{j}} \exp(E_{\hat{j}}|t|) , \tag{5.7.19a}$$

then

$$|j_B(t)| \leq \int_o^{|t|} C_{\hat{j}} \exp(E_{\hat{j}} t_1) dt_1 \leq C_{\hat{j}} \cdot |t| \cdot \exp(E_{\hat{j}}|t|) .$$

To get the bounds $C_{\hat{\jmath}}$, $E_{\hat{\jmath}}$ we note that

$$\hat{\jmath}_B(t) = \frac{1}{4\pi i} \oint e^{t/z} \hat{\jmath}(z) dz/z^2 ,$$

$$\partial D_2(0, \tfrac{4}{\kappa} \rho_\Gamma^2)$$

and it follows that

$$C_{\hat{\jmath}} = \sup_{z \in D_2(0, \frac{4}{\kappa} \rho_\Gamma^2)} \left| \frac{\hat{\jmath}(z)}{z} \right| ,$$

and furthermore (L35-36),

$$E_{\hat{\jmath}} = \frac{\kappa}{4\rho_\Gamma^2} \equiv sEJH .$$ (5.7.20)

The calculation of an upper bound for $C_{\hat{\jmath}}$ is again standard. We have, (L37-38),

$$C_{\hat{\jmath}} \leq (E_{\hat{\jmath}})^2 \cdot |J^u(\rho^+ z/\rho_\Gamma)|_B \equiv sCJH .$$ (5.7.21)

(We should calculate in fact $|\frac{1}{z^4} J^u(\rho^+ z/\rho_\Gamma)|_B$ and not $|J^u(\rho^+ z/\rho_\Gamma)|$, but it is easy to check that the function r1BOUND yields the same answer in both cases.)

Let now $g(z) = \frac{4}{\kappa}(\Gamma((\frac{\kappa}{4}z)^{1/2}))^2$. g is analytic on $D_2(0, \frac{4}{\kappa} \rho_\Gamma^2)$, and we need to bound the Borel transform of $Y(z) = \frac{1}{g(z)} - \frac{1}{z} - 1$. Following exactly the procedure we have used to calculate (5.7.4) and (5.7.8) we get, (L42),

$$E_\gamma = \kappa/4\rho_\Gamma^2 \equiv \text{sEI2ETA} \quad (\equiv \text{sEJH}) \; , \tag{5.7.22}$$

and (L43-49)

$$C_\gamma \le (E_\gamma)^2 \cdot |(1/H^u(z))^2 - 1 - \frac{4}{\kappa}\rho_\Gamma^2 z^2|_1 \equiv \text{sCI2ETA} \; . \tag{5.7.23}$$

Let now $f(z) = F((\frac{\kappa}{4}z)^{1/2})$ be the normal form of the function $F(z)$ and let f_B be the Borel transform of f. In principle we could use the bounds (2.4.17), (2.4.18) to give a bound for the function f_B at hand. However since we establish in the present section bounds only on the domain S_2 (see Fig. 5.7.1), we can sharpen (2.4.15), (2.4.16) using Lemma 5.7.1. We replace (2.4.15) by the bound

$$\frac{C_3}{1+C_3} \cdot \sum_{m=1}^{\infty} (C_\gamma(1 + C_3))^m \frac{|t|^{m+1}}{(m+1)!}$$

$$= \frac{C_3}{C_\gamma(1+C_3)^2} \cdot \{\exp(C_\gamma(1 + C_3)|t|) - 1 - C_\gamma(1 + C_3)|t|\}$$

$$\le \frac{C_3}{C_\gamma(1+C_3)^2} \cdot \{\exp(C_\gamma(1 + C_3)|t|) - 1\}$$

and get instead of (2.4.16)

$$|f_B(t)| \le C_3\tilde{C}_{\hat{j}} e^{(\tilde{E}_{\hat{j}}+1)|t|}$$

$$+ \frac{C_3 C_{\hat{j}}}{C_\gamma(1+C_3)^2} e^{(E_{\hat{j}}+C_\gamma(1+C_3))|t|} |E_\alpha(t)|_{\alpha=C_\gamma(1+C_3)}$$

and where $\tilde{E}_{\hat{j}}$ is chosen such as to optimize the bound, namely

(L53-54)

$$\tilde{E}_{\hat{J}} = E_{\hat{J}} + C_\gamma (1 + C_3) - 1 \equiv sEJHT \ , \qquad (5.7.24)$$

and hence $\tilde{C}_{\hat{J}}$ is bounded by (L55-58)

$$\tilde{C}_{\hat{J}} = (\tilde{E}_{\hat{J}})^2 |J^u(\tilde{\rho}z/\rho_\Gamma)|_B \equiv sCJHT \qquad (5.7.25)$$

with (L55),

$$\tilde{\rho} = (\kappa/4\tilde{E}_{\hat{J}})^{1/2} \ .$$

Using now Lemma 5.7.1 for the bound on E_α we get for $t \in S_2$

$$|f_B(t)| \le C_f \cdot \exp(E_f|t|) \ ,$$

where (L68-71)

$$C_f = C_3\tilde{C}_{\hat{J}} + \frac{C_3 C_{\hat{J}}}{C_\gamma (1+C_3)^2}(1 + C_3(1 + C_\gamma(1 + C_3))) \equiv sCF\emptyset \ , \qquad (5.7.26)$$

and (L66)

$$E_f = E_{\hat{J}} + C_\gamma (1 + C_3) \equiv sEF\emptyset \ . \qquad (5.7.27)$$

(In L64 we check that $C_\gamma(1 + C_3) > 1$, i.e. that $\tilde{E}_{\hat{J}} > E_{\hat{J}}$. This is a little bit artificial since in the case where this inequality is not satisfied, an error would already have occured in L56 since $\tilde{\rho}$ would turn out to be larger than ρ_Γ. We spread such checks into the program in order to alert the reader that something needs to be checked at these places.)

This completes our discussion of the subroutines CREDUCE and FREDUCE. The next step is to explain the subroutine CxFØ.

```
      SUBROUTINE CxFO(xGAMMA,xJ,xFO,IDEG)
      INCLUDE BIGN.STUFF
      COMMON/DOMAIN/sRHOGO,sRHOGAMMA,sXOh,sCENTh,sRHOh,sCENTG,sRHOG
     *           ,sRHOhB,sRHOPhB,sXOhB,sPHIO,sxKAPPA,sKAPPA
     *           ,sRHOP,sRHO1
      DIMENSION xGAMMA(1),xJ(1),xFO(1)
      DIMENSION xT1(NDEG)

C  Solve Ecalle's equation iteratively
C  ===================================
      NSAVE=N
      CALL FSETZERO(xFO)
      DO 1 I=3,IDEG-1
      CALL DEG(I+1)
      CALL CCOMP(xFO,xGAMMA,xT1)
      CALL FMINUS(xFO,xT1)
      CALL FMINUS(xJ,xT1)
      sIM1=siCONST((I-1)*2)
      xFO(I)=sPROD(xT1(I+2),sQUOT(sxKAPPA,sIM1))
    1 CONTINUE
      CALL DEG(NSAVE)

      RETURN
      END
```

This subroutine is essentially self-explaining. We have
$\hat{\Gamma}(x) = x - \frac{2}{\kappa} x^3 + O(x^4)$ and $J_{P,thin}(x) = O(x^4)$ and therefore
$\hat{F}(x) = O(x^2)$. Suppose $\hat{F}$ is known up to order x^{M-1} and call
$\hat{F}_{M-1}(x) = \sum_{n=2}^{M-1} F_n \cdot x^n$, then the coefficient of x^M, F_M is given by
the equation

$$F_M \cdot [(x - \frac{2}{\kappa} x^3 + O(x^4))^M - x^M]_{(M+2)} =$$

$$= [J_{P,thin}(x) - (\hat{F}_{M-1}(\hat{\Gamma}(x)) - \hat{F}_{M-1}(x))]_{(M+2)} \quad ,$$

where [ ]$_{(M+2)}$ means that we extract the coefficient of the power x^{M+2} of the expression in the bracket [ ]. It follows that

$$F_M = [J_{P,thin}(x) - (\hat{F}_{M-1}(\hat{\Gamma}(x)) - \hat{F}_{M-1}(x))]_{M+2}/(-\frac{2M}{\kappa}) \ .$$

This is exactly what the subroutine CxFØ calculates for M = 2 up to M = IDEG - 2, and we call the subroutine CxFØ with IDEG = N in L268 of FvKI. Therefore we know the coefficients of $\hat{F}(x)$ up to order N - 2. This will have to be remembered in the discussion of the subroutine FSUMBOREL which we explain next.

5.7.4 Constructive version of Loeffel's method for Borel summation

FSUMBOREL

```
 1:          SUBROUTINE FSUMBOREL(xF0,IDEG,sC,sR,vhB)
 2:          INCLUDE BIGN.STUFF
 3:          COMMON/DOMAIN/sRHOG0,sRHOGAMMA,sX0h,sCENTh,sRHOh,sCENTG,sRHOG
 4:      *                ,sRHOhB,sRHOPhB,sX0hB,sPHI0,sxKAPPA,sKAPPA
 5:      *                ,sRHOP,sRHO1
 6:          COMMON/SCALE /sX00
 7:          COMMON/LOFFEL/sBETA,xPHI(NDEG),vINT1(NHALF,NDEG)
 8:      *                ,vINT2(NHALF,NDEG)
 9:          DIMENSION xF0(1),vhB(1)
10:          DIMENSION xFB1(NDEG),xFB2(NDEG)
11:          DIMENSION vT1(NDEG)
12:
13:C   Make coefficients of Borel function
14:C   ===================================
15:
16:          CALL FSETZERO(xFB1)
17:          sP1=sONE
18:          IK=1
19:          I=1
20:          IB1=0
21:    1 CONTINUE
22:          IB1=IB1+1
23:          xFB1(IB1)=sQUOT(xF0(I+2),sP1)
24:          sP1=sPROD(sP1,siCONST(IK))
25:          IK=IK+1
26:          I=I+2
27:          IF(I+2.LE.IDEG-1)GOTO 1
```

```
28:          IDEG1=IB1-1
29:          CALL FSETZERO(xFB2)
30:          sP1=sONE
31:          IK=1
32:          I=2
33:          IB2=0
34:    2 CONTINUE
35:          IB2=IB2+1
36:          xFB2(IB2)=sNEG(sQUOT(xF0(I+2),sP1))
37:          sP1=sPROD(sP1,siCONST(IK))
38:          IK=IK+1
39:          I=I+2
40:          IF(I+2.LE.IDEG-1)GOTO 2
41:          IDEG2=IB2-1
42:
43:
44:C Rearrange coefficients by conformal mapping
45:C =============================================
46:
47:          NSAVE=N
48:          CALL DEG(IDEG1)
49:          CALL CCOMP(xFB1,xPHI,xFB1)
50:          CALL DEG(IDEG2)
51:          CALL CCOMP(xFB2,xPHI,xFB2)
52:          CALL DEG(NSAVE)
53:
54:
55:C Construction of vhB: sum polynomial part of Borel function
56:C ==========================================================
57:
58:          DO 3 IK=1,IDEG1+1
59:          DO 4 IN=1,NERR
60:          vT1(IN)=vINT1(IDEG1+2-IK,IN)
61:    4 CONTINUE
62:          IF(IK.EQ.1)CALL FsMULT(vT1,xFB1(IDEG1+2-IK),vhB)
63:          IF(IK.GT.1)CALL FsMULADD(vT1,xFB1(IDEG1+2-IK),vhB)
64:    3 CONTINUE
65:          DO 5 IK=1,IDEG2+1
66:          DO 6 IN=1,NERR
67:          vT1(IN)=vINT2(IDEG2+2-IK,IN)
68:    6 CONTINUE
69:          CALL FsMULADD(vT1,xFB2(IDEG2+2-IK),vhB)
70:    5 CONTINUE
71:C..normalize vhB
72:          vhB(1)=sZERO
73:
74:
```

```
75:C   Construction of vhB: complete error term
76:C   ==========================================
77:
78:C   Calculate worst point in domain
79:C   -------------------------------
80:        sZ=sPOWER(sSUM(sXOhB,sRHOPhB),2)
81:
82:C   Calculate bounds from even part
83:C   -------------------------------
84:        sT1=sHIGH(sC,sR,sZ,IDEG1,1)
85:
86:C   Calculate bounds from odd part
87:C   ------------------------------
88:        sT2=sHIGH(sC,sR,sZ,IDEG2,2)
89:
90:C   Add bounds and calculate L1-NORM
91:C   --------------------------------
92:        sT1=sSUM(sT1,sT2)
93:C..sT1 contains sup(D(0,rhop)) of contribution to general term;
94:C   L1-norm is bounded by sT1*(rho/rhop)/(1-(rho/rhop))
95:        sT2=sQUOT(sRHOhB,sRHOPhB)
96:        rT1=rMAXABS(sPROD(sT1,sQUOT(sT2,sDIFF(sONE,sT2))))
97:        vhB(NERR)=error(rT1,rH(vhB))
98:        RETURN
99:        END
```

This subroutine is essentially the implementation of Loeffel's

method as it is described in Section 3. Let again

$f(z) = F((\frac{\kappa}{4}z)^{1/2})$ be the normal form of the function F and f_B

its Borel transform. Since f satisfies

$$f(z) = \int_0^\infty e^{-t/z} f_B(t)\, dt$$

it follows by a change of variables that the function

$f_\mathcal{L}(z) = f(\frac{4}{\kappa}z) = F(z^{1/2})$ is the inverse Borel transform of the

function $f_{B,\mathcal{L}}(t) = \frac{4}{\kappa} f_B(\frac{4}{\kappa}t)$, i.e.

$$f_\mathcal{L}(z) = \int_0^\infty e^{-t/z} f_{B,\mathcal{L}}(t)\, dt \; .$$

As announced at the beginning of this section it is to $f_{B,\mathcal{L}}$ and

not to f_B that we want to apply Loeffel's method since the coef-

ficients of $f_{B,\mathcal{L}}$ can be calculated with higher precision. (The

factor $\frac{4}{\kappa}$ in the relation between f and F has been avoided by passing to $f_\mathcal{L}(z) = F(z^{1/2}).)$ Let now

$$\hat{F}(x) = \sum_{k=2}^{\infty} F_k x^k ,$$

then we get

$$\hat{f}_\mathcal{L}(x) = \sum_{k=1}^{\infty} f_k x^{(k+1)/2} ,$$

where $f_k = F_{k+1}$ and $\hat{f}_\mathcal{L}$ is the formal power series (in $x^{1/2}$) and corresponding to the function $f_\mathcal{L}$. By definition the Taylor expansion of $f_{B,\mathcal{L}}(t)$ at $t = 0$ is given by

$$f_{B,\mathcal{L}}(t) = \sum_{k=1}^{\infty} \frac{f_k}{((k-1)/2)!} t^{(k-1)/2} , \tag{5.7.28}$$

where $f_{B,\mathcal{L}}(t)$ is by construction analytic in $S_{2,\beta}$ with $\beta \leq \frac{\kappa}{2}$ and satisfies there the bound

$$|f_{B,\mathcal{L}}(t)| \leq C_\mathcal{L} \exp(|t|/R_\mathcal{L}) ,$$

where $C_\mathcal{L} = \frac{4}{\kappa} C_f \equiv sC$ and $R_\mathcal{L} = (\frac{4}{\kappa} E_f)^{-1} \equiv sR$. (See L287-288 of FvKI.) Next we define as in Section 3 the two functions $f_{B,\mathcal{L},1}$ and $f_{B,\mathcal{L},2}$, which are analytic on $S_\beta = \{-\beta \log(1-t) | t \in \mathbb{C}, |t| < 1\}$ and have at $t = 0$ the expansions

$$f_{B,\mathcal{L},1}(t) = \sum_{k=0}^{\infty} \frac{f_{2k+1}}{k!} t^k \equiv xFB1 \tag{5.7.29}$$

(L16-28)   and furthermore

$$f_{B,\mathcal{L},2}(t) = -\sum_{k=1}^{\infty} \frac{f_{2k}}{(k-\frac{1}{2})!} t^k .$$

(5.7.30)

(For historical reasons we have chosen a minus sign in the definition of (5.7.30).) To $f_{B,\mathcal{L},1}$ one could now immediately apply
Loeffel's resummation procedure, but to $f_{B,\mathcal{L},2}$ we must first
apply the transformation (3.12) and get a function $\tilde{f}_{B,\mathcal{L},2}$ which
is again analytic on S_β and has the expansion (see Section 3)

$$\tilde{f}_{B,\mathcal{L},2}(t) = -\sum_{k=1}^{\infty} \frac{f_{2k}}{(k-\frac{1}{2})!} \cdot (\frac{(k-\frac{1}{2})!}{(k-1)!}) t^{k-1}$$

$$= -\sum_{k=1}^{\infty} \frac{f_{2k}}{(k-1)!} t^{k-1} \equiv xFB2$$

(5.7.31)

(L29-41).   Remember that the coefficients of $\hat{F}(x) \equiv xF\emptyset$ are known
up to degree $N - 2$. Therefore the coefficients of $f_{B,\mathcal{L},1}$ are
known up to degree $[(N-4)/2] \equiv IDEG1$ and those of $\tilde{f}_{B,\mathcal{L},2}$ up to
degree $[(N-3)/2 - 1] \equiv IDEG2$. According to what has been explained
in Section 3 we now have to rearrange the coefficients of $f_{B,\mathcal{L},2}$
in order to get the functions $\hat{f}_{B,\mathcal{L},1}(t)$ and $\hat{f}_{B,\mathcal{L},2}(t)$ which
are analytic on the unit disk. This is done by substituting the
formal power series $\hat{\varphi}_\beta(x)$ of the function $\varphi_\beta(t) = -\beta \cdot \log(1 - t)$.
This we do in L48-49 and L51-52. $\hat{\varphi}_\beta(x) \equiv xPHI$ is constructed
in the subroutine CSETPHI which has already been called as a
preparatory step right at the beginning of the main program. (L66-
68 of the main program.) We give the listing of the subroutine
CSETPHI for completeness.

```
      SUBROUTINE CSETPHI(x)
      INCLUDE BIGN.STUFF
      COMMON/LOFFEL/sBETA,xPHI(NDEG),vINT1(NHALF,NDEG)
     *              ,vINT2(NHALF,NDEG)
      DIMENSION x(1)


C  x(z)=sBETA*(-log(1-z))
C  =======================

      CALL CSETLOG(x)
      CALL FsMULT(x,sBETA,x)


      RETURN
      END
```

We now have constructed the coefficients of $\hat{f}_{B,\mathcal{L},1}$ and $\hat{f}_{B,\mathcal{L},2}$ and want now to construct the functions $f_{\mathcal{L},1}(z)$ and $f_{\mathcal{L},2}(z)$ ((3.7) and (3.13) of Section 3). To achieve that, we need to construct the functions $m_{1,\mathcal{L}}(z)$ and $\tilde{m}_{2,\mathcal{L}}(z)$, but since we need afterwards the function $F(z) = f(z^2)$ and not $f(z)$ itself we construct directly $M_{1,\mathcal{L}}(z) = m_{1,\mathcal{L}}(z^2)$ and $M_{2,\mathcal{L}}(z) = \tilde{m}_{2,\mathcal{L}}(z^2)$. This is done in the subroutine PREPBOREL which we now explain in short.

```
      SUBROUTINE PREPBOREL(sXOhB,sRHOhB,IDEG)
      INCLUDE BIGN.STUFF
      COMMON/LOFFEL/sBETA,xPHI(NDEG),vINT1(NHALF,NDEG)
     *              ,vINT2(NHALF,NDEG)
      DIMENSION vT1(NDEG),vT2(NDEG),vT3(NDEG)


C  Compute functions M1,k for k=0,1,... and M2,k for k=1,2,...
C  ==========================================================

C  Case of functios M1,k
C  ---------------------
C..vT1=(rhoB*z-xOhB)**2
      CALL FSETZERO(vT1)
      vT1(1)=sPROD(sXOhB,sXOhB)
      vT1(2)=sNEG(sPROD(sPROD(sTWO,sXOhB),sRHOhB))
      vT1(3)=sPROD(sRHOhB,sRHOhB)
      IK=0
      DO 1 IN=1,NERR
      vINT1(IK+1,IN)=vT1(IN)
    1 CONTINUE
C..vT1 is used as accumulator below
      CALL FINVERS(vT1,vT2)
      CALL FsMULT(vT2,sBETA,vT2)
      sFACT=sONE
C..start loop
      DO 2 II=3,IDEG-3,2
      IK=IK+1
      sFACT=sPROD(sFACT,siCONST(IK))
      vT2(1)=sSUM(vT2(1),sONE)
      CALL FINVERS(vT2,vT3)
      CALL FMULT(vT1,vT3,vT1)
      CALL FsMULT(vT1,sFACT,vT3)
      DO 3 IN=1,NERR
      vINT1(IK+1,IN)=vT3(IN)
    3 CONTINUE
    2 CONTINUE

C  Case of functions M2,k
C  ----------------------
C..vT1=(sXOhB-rhoB*Z)
      CALL FSETLIN(vT1,sXOhB,sNEG(sRHOhB))
      IK=-1
C..start loop
      DO 4 II=2,IDEG-3,2
      IK=IK+1
      DO 5 IN=1,NERR
      vT2(IN)=vINT1(IK+1,IN)
    5 CONTINUE
      CALL FMULT(vT1,vT2,vT2)
      DO 6 IN=1,NERR
      vINT2(IK+1,IN)=vT2(IN)
    6 CONTINUE
    4 CONTINUE

      RETURN
      END
```

We need to construct the functions

$$M_{1,k}(z) = \beta \cdot \frac{k!}{\prod\limits_{j=0}^{k} (\frac{\beta}{z^2}+j)} \equiv vINT1[k+1,] \tag{5.7.32}$$

for $k = 0,1,2,\ldots$ IDEG1 (see (3.8)), for $z \in D(-x_{oh_B}, \rho_{h_B})$, and the functions

$$M_{2,k}(z) = (-z)M_{1,k}(z) \equiv vINT2[k+1,] , \tag{5.7.33}$$

for $k = 0,1,2,\ldots$ IDEG2, again for $z \in D(-x_{oh_B}, \rho_{h_B})$. (The minus sign in the definition of $M_{2,k}$ compensates that in the defini- tions (5.7.30) and (5.7.31). We have chosen to put these minus signs because with our definition of the square root we have $\sqrt{z^2} = -z$ if Rez < 0.) The construction of the vectors vINT1 and vINT2 is completely trivial given the subroutine package of Section 4, and the program of the subroutine PREPBOREL is there- fore selfexplaining.

Given the functions $M_{1,k}$ and $M_{2,k}$ we can now continue our discussion of the subroutine FSUMBOREL. In L58-72 we sum the vectors vINT1 [i,] and vINT2 [i,] with their weights xFB1 [i], xFB2[i] into the vector vhB $\equiv h_B^u$. It remains the higher order contributions (i > IDEG1), (i > IDEG2) to be added. The necessary bounds have been established in (3.11) and (3.19) and are calculated in L84 and L88 using the functions sHIGH(,,,,1) and sHIGH(,,,,2) respectively. sHIGH establishes these bounds on $D(-x_{oh_B}, \rho_{h_B}^+)$ and we use then the inequality

$$\left| h\left(\frac{\rho h_B}{\rho h_B^+}z\right) - h(0)\right|_1 \leq \frac{(\rho h_B/\rho h_B^+)}{1-(\rho h_B/\rho h_B^+)} \cdot \sup_{z\in D(0,\rho h_B^+)} |h(z)| \,, \qquad (5.7.34)$$

(which is in particular true for a function $h(z)$ with $h(0) = 0$),

to establish the bound on the ℓ^1-norm on the domain $D(-x_{oh_B}, \rho h_B)$

(L92-97). Note that the bound (3.11) is a increasing function of

Rez. Therefore it suffices to establish (3.11) for the worst point

in the domain in question which in the case at hand is

Rez = $(x_{oh_B} + \rho h_B^+)^2 \equiv sZ$ (see L80). This completes our discussion

of the subroutine FSUMBOREL up to the discussion of the function

sHIGH which we explain next.

```
 1:        COMPLEX FUNCTION sHIGH*16(sC,sR,sZ,IDEG,IEO)
 2:        INCLUDE BIGN.STUFF
 3:        COMMON/LOFFEL/sBETA,xPHI(NDEG),vINT1(NHALF,NDEG)
 4:       *              ,vINT2(NHALF,NDEG)
 5:
 6:
 7:C  Calculate contribution of higher order of xFO to rG(vhB)
 8:C  =======================================================
 9:
10:C  Add contributions of terms KMIN1+1,KMIN2+1...MBIG
11:C  -------------------------------------------------
12:        sARG=sQUOT(sBETA,sR)
13:        sARG2=sQUOT(sBETA,sZ)
14:        sT2=sZ
15:        sT3=sARG2
16:        DO 1 I=1,IDEG+1
17:        sT3=sSUM(sT3,sONE)
18:        sT2=sPROD(sT2,sQUOT(siCONST(I),sT3))
19:      1 CONTINUE
20:        IF(IEO.EQ.2)THEN
21:          sT2=sPROD(sT2,sSQRT(sZ))
22:        ENDIF
23:        sT1=sZERO
24:        MBIG=MAXO(100,IDEG+1)
25:        DO 2 IK=IDEG+2,MBIG
26:        sT3=sSUM(sT3,sONE)
27:        sT2=sPROD(sT2,sQUOT(siCONST(IK),sT3))
28:        sT5=siCONST(IK)
29:        IF(IEO.EQ.1)THEN
30:          sT6=sSUM(sONE,sQUOT(sT5,sARG))
31:          sT6=sTOs(sT6,sARG)
32:          sT7=sSUM(sONE,sQUOT(sARG,sT5))
33:          sT7=sTOs(sT7,sT5)
34:          sT4=sPROD(sT6,sT7)
35:        ENDIF
36:        IF(IEO.EQ.2)THEN
```

```
37:          sTRH=sTRHOPTIMAL(sARG,IK)
38:          sTR=sPROD(sQUOT(sSUM(sT5,sONE),sSUM(sT5,sTWO)),sTRH)
39:          sT6=sINV(sDIFF(sONE,sTRH))
40:          sT6=sTOs(sT6,sARG)
41:          sT7=sINV(sTRH)
42:          sT7=sTOs(sT7,sT5)
43:          sT4=sPROD(sT6,sT7)
44:          sT6=sQUOT(sPROD(sSUM(sONE,sTR),sTRH)
45:     *        ,sPROD(sSUM(sTRH,sTR),sDIFF(sTRH,sTR)))
46:          sT7=sSQRT(sNEG(sLOG(sDIFF(sONE,sTRH))))
47:          sT8=sTOs(sQUOT(sSUM(sT5,sTWO),sSUM(sT5,sONE)),sT5)
48:          sT4=sPROD(sT4,sPROD(sSQRT(sQUOT(sPI,sBETA)),sPROD(sT6
49:     *        ,sPROD(sT7,sT8))))
50:      ENDIF
51:      sT1=sSUM(sT1,sPROD(sT2,sT4))
52:    2 CONTINUE
53:      sT1=sPROD(sT1,sC)
54:
55:C   Integral estimate of contribution of terms M+1...
56:C   --------------------------------------------------
57:      IF(IEO.EQ.1)THEN
58:        sCP=sC
59:        sRP=sR
60:      ENDIF
61:      IF(IEO.EQ.2)THEN
62:C..choose optimal constant
63:        sT3=saCONST(.4D+00)
64:        sT4=sSUM(sQUOT(sBETA,sR),sQUOT(sBETA,sPROD(sTWO
65:     *        ,sPROD(sT3,sEXP(sONE)))))
66:C..calculate bound
67:        sCP=sPROD(sSUM(sXOhB,sRHOPhB),sPROD(sPROD(sTWO,sSQRT(sPI))
68:     *        ,sPROD(sINV(sBETA),sPROD(sSQRT(sT3)
69:     *        ,sPROD(sC,sPROD(sSUM(sONE,sT4),sEXP(sNEG(sPROD(sT4
70:     *        ,sLOG(sQUOT(sT4,sSUM(sONE,sT4)))))))))))))
71:        sRP=sINV(sSUM(sINV(sR),sSUM(sINV(sPROD(sTWO,sPROD(sT3
72:     *        ,sEXP(sONE)))),sINV(sBETA))))
73:      ENDIF
74:      sARG=sQUOT(sBETA,sRP)
75:      sM=siCONST(MBIG)
76:      sT2=sTOs(sSUM(sINV(sARG),sINV(sM)),sARG)
77:      sT3=sEXP(sARG)
78:      sT5=sPROD(sZ,sSQRT(sPROD(sPI,sTWO)))
79:      sT6=sSQRT(sINV(sARG2))
80:      sT7=sINV(sTOs(sSUM(sONE,sQUOT(sARG2,sM)),sM))
81:C..sT8 has too large exponent. Multiply below twice with sT8**.5
82:      sT8=sTOs(sARG2,sQUOT(sARG2,sTWO))
83:      sT9=sDIFF(sPROD(siCONST(12),sARG2),sONE)
84:      IF(DREAL(sT9).LT.rZERO)WRITE(6,*)'error in FSUMBOREL, 1)'
85:      sT9=sSUM(sONE,sINV(sT9))
86:      sT10=sDIFF(sPROD(siCONST(12),sM),sONE)
87:      IF(DREAL(sT10).LT.rZERO)WRITE(6,*)'error in FSUMBOREL, 2)'
88:      sT10=sSUM(sONE,sINV(sT10))
89:      sARG3=sDIFF(sARG2,sSUM(sARG,sONE))
90:      IF(DREAL(sARG3).LT.rZERO)WRITE(6,*)'error in FSUMBOREL, 3)'
91:      sT11=sINV(sARG3)
92:      sT12=sTOs(sM,sNEG(sARG3))
93:      sT12=sPROD(sT12,sT8)
94:C..multiply with sCP
95:      sT12=sPROD(sCP,sT12)
96:C..add contributions
97:      sHIGH=sSUM(sT1,sPROD(sPROD(sPROD(sPROD(sPROD(sPROD(sPROD
98:     *        (sPROD(sT2,sT3),sT5),sT6),sT7),sT8),sT9),sT10)
99:     *        ,sT11),sT12))
100:
101:
102:      RETURN
103:      END
```

We first explain the case IEO = 1. The function sHIGH is then just a straightforward implementation of the bound (3.11) (as a function of z^2 instead of z since we are working with $M_{1,k}(z) = m_{1,k}(z^2))$, with C ≡ sC, R ≡ sR, β ≡ sBETA and $Rez^2 \equiv sZ$. However in order to gain precision we do not apply (3.11) to all k with k > IDEG1 but prefer to sum first explicitly the products of a bound on (3.8) times the bound (3.9) for k = IDEG1 + 1 up to MBIG = 100. We use then (3.11) only to bound the sum of the remaining terms, i.e. we use (3.11) with M ≡ MBIG. Note that it follows from (3.8) that $|m_{1,k}(z)|$ is an increasing function of Rez, so that we get

$$|M_{1,k}(z)| \leq \beta \cdot \frac{k'}{\prod\limits_{j=0}^{k} (\frac{\beta}{Rez^2}+j)} \ . \tag{5.7.35}$$

In L12-19 we calculate the bound (5.7.35) for k = IDEG1. In L23-35 and L51-53 we sum bounds as described, implementing in addition to (5.7.35) the bound

$$|\hat{f}_{B,\mathcal{L},1}^{(k)}| \leq C \cdot (1 - \rho)^{-\beta/R} \cdot \rho^{-k} \ , \tag{5.7.36}$$

where

$$\dot{\rho} = \frac{k}{k+\beta/R} = 1 - \frac{\beta/R}{k+\beta/R} \ . \tag{5.7.37}$$

L57-60 and L74-99 finally contain the implementation of (3.11). We have added several tests to make sure that all the denominators in (3.11) are positive. This proves that Loeffel's resummed series converges absolutely in the prescribed domains. (Loeffel's sum does not converge absolutely in the whole domain of convergence of the Borel sum.) The case IEO = 2 necessitates some minor changes of the bounds, but again we first sum explicitly the products of a bound on $M_{2,k}(z)$ times the bound (3.18) and use afterwards a bound analogous to (3.11) to bound the remaining terms. We have

$$|M_{2,k}(z)| \le |z||M_{1,k}(z)| \, , \qquad (5.7.38)$$

and we can estimate on our domain $|z|$ by $(Rez^2)^{1/2} \equiv sSQRT(sZ)$. We therefore multiply in L20-22 the bound (5.7.35) with sSQRT(sZ) and get the bound (5.7.38). The bound (5.7.36) is replaced by what we get from (3.19) and (3.4), namely

$$|\hat{f}_{B,\mathcal{L},2}^{(k)}| \le \frac{1}{2}(\frac{\pi}{\beta})^{1/2} \cdot (\frac{\hat{\rho}}{\rho})^k \cdot \frac{(1+\rho)}{(\hat{\rho}-\rho)} (-\log(1-\hat{\rho}))^{1/2} \cdot (\frac{1}{\rho})^k$$

$$\cdot \exp(-\frac{\beta}{R} \cdot \log(1-\hat{\rho})) \, , \qquad (5.7.39)$$

where $0 < \rho \le \hat{\rho} < 1$, but otherwise arbitrary. We choose here first

$$\rho \equiv \rho(\hat{\rho}) = (\frac{k+1}{k+2})\hat{\rho} \, ,$$

and optimize then numerically the expression

$$\frac{(1+\rho)\hat{\rho}}{(\hat{\rho}^2-\rho^2)} (-\log(1-\hat{\rho}))^{1/2} \cdot \exp(-\frac{\beta}{R} \cdot \log(1-\hat{\rho})) \cdot (\frac{1}{\rho})^k \qquad (5.7.40)$$

using the Newton algorithm. This we do in the function sTRHOPTIMAL which calls in turn the function aFTRH in order to evaluate

(5.7.40). All this is carried out in L36-50 of sHIGH and re-
places L29-35 of before. For sake of completeness we give the
listings of the functions sTRHOPTIMAL and aFTRH at this point.

```
COMPLEX FUNCTION sTRHOPTIMAL*16(sARG,K)
INCLUDE BIGN.STUFF
aARG=aCENT(sARG)
aTRH=K/(aARG+K+1)
aT1=1.D-8
DO 1 I=1,5
aTF =aFTRH(aTRH,aARG,K)
aTF1=aFTRH(aTRH+aT1,aARG,K)
aTF2=aFTRH(aTRH+2*aT1,aARG,K)
aTRH=aTRH-aT1*(aTF1-aTF)/(aTF2-2*aTF1+aTF)
1 CONTINUE
sTRHOPTIMAL=saCONST(aTRH)
RETURN
END
```

```
REAL FUNCTION aFTRH*8(aTRH,aARG,K)
INCLUDE BIGN.STUFF
aTR=((K+rONE)/(K+rTWO))*aTRH
aFTRH=(((1+aTR)*aTRH*DSQRT(-DLOG(1-aTRH)))/((aTR+aTRH)
*      *(aTRH-aTR)))*DEXP(-aARG*DLOG(1-aTRH)-K*DLOG(aTRH))
RETURN
END
```

For the contributions from $k > M$ we proceed as follows. Again
we bound $M_{2,\mathcal{L}}(z)$ by $|z| \cdot M_{1,\mathcal{L}}(z)$, i.e. we use the bound (3.10)
and multiply it with $(x_{oh_B} + \rho_{h_B}^+)$. Furthermore we want to bring
the bound (3.19) into the form (3.9) such that we can use again
(3.11) changing nothing but the values of C and R in that
bound. We have (see (3.19))

$$|\hat{f}_{B,\mathcal{L},2}^{(\mathcal{L})}| \leq \pi^{1/2} \cdot \frac{1}{\beta} \cdot \frac{1}{\hat{\rho}-\rho} \cdot C \cdot a \cdot (\frac{1}{\rho})^k \cdot (\frac{1}{1-\rho})^{\beta/R''} , \tag{5.7.41}$$

where $\dfrac{1}{R''} = \dfrac{1}{R} + \dfrac{1}{2a^2e}$, and choosing $\hat{\rho} = (1 + \dfrac{\beta}{R''}\rho)/(1 + \dfrac{\beta}{R''})$ as described for (3.19) we get

$$\hat{\rho} - \rho = \frac{1-\rho}{1+\beta/R''} \ ,$$

$$1 - \hat{\rho} = \frac{\beta/R''}{1+\beta/R''}(1 - \rho) \ ,$$

and therefore

$$\left| f_{B,\ell,2}^{(\ell)} \right| \leq \pi^{1/2} \cdot \frac{1}{\beta} \cdot (1 + \beta/R'') \cdot \left(\frac{\beta/R''}{1+\beta/R''}\right)^{-\beta/R''} \cdot a$$

$$\cdot \frac{C}{(1-\rho)^{\beta/R''+1}\rho^k} \tag{5.7.42}$$

and this is exactly of the form (3.9) and takes its minimal value for $\rho = 1 - (\beta/R'')/(k + (\beta/R''))$. From this and the previous remark on $M_{2,k}$ it follows that we can use (3.11) to bound the remaining terms, if we replace C by (L63-65)

$$C' = \pi^{1/2} \cdot \frac{1}{\beta} \cdot 2 \cdot (1+\frac{\beta}{R''}) \left(\frac{\beta/R''}{1+\beta/R''}\right)^{\beta/R''} \cdot C \cdot a \cdot (x_{ohB} + \overset{+}{\rho} h_B)$$

and R by (L67-75)

$$R' = \frac{1}{R''} + \frac{1}{\beta} = \frac{1}{R} + \frac{1}{2a^2e} + \frac{1}{\beta} \ .$$

We choose $a^2 = 0.4 \equiv sT3$ (L63). Furthermore $\dfrac{\beta}{R''} \equiv sT4$ (L64-65).

This completes our discussion of the subroutine sHIGH and of Loeffel's method of constructive Borel summation.

5.7.5 Ecalle's equation for general inhomogeneities

We next discuss the subroutine FADDGENERAL, where we complete the Borel summation procedure for <u>vectors</u> by adding bounds on ΔF (see Table 5.7.1)

```
 1:        SUBROUTINE FADDGENERAL(vhB,sGJ,sHJ,sC,sR,sGGAMMA,sHGAMMA
 2:       *                     ,sZ,sETA,sL5,sL6,vhBP)
 3:        INCLUDE BIGN.STUFF
 4:        COMMON/DOMAIN/sRHOGO,sRHOGAMMA,sXOh,sCENTh,sRHOh,sCENTG,sRHOG
 5:       *             ,sRHOhB,sRHOPhB,sXOhB,sPHIO,sxKAPPA,sKAPPA
 6:       *             ,sRHOP,sRHO1
 7:        DIMENSION vhB(1),vhBP(1)
 8:
 9:
10:C  Store Borel part in result
11:C  =========================
12:
13:        CALL FEQUAL(vhB,vhBP)
14:
15:
16:C  Calculate bounds on image
17:C  =========================
18:
19:        CALL FBNDIM(sC,sR,sGGAMMA,sHGAMMA,sZ,sETA,sGJFO,sHJFO)
20:
21:
22:C  Calculate total image
23:C  =====================
24:
25:        sGJTOT=sUPPER(sSUM(sGJ,sGJFO))
26:        sHJTOT=sUPPER(sSUM(sHJ,sHJFO))
27:
28:
29:C  Calculate bound on preimage
30:C  ===========================
31:
32:        CALL FBNDPREIM(sGJTOT,sHJTOT,sRHOP,sL5,sL6,sGFO,sHFO)
33:
34:
35:C  Estimate contribution to L1-norm of vhB
36:C  =======================================
37:
38:        sT1=sQUOT(sRHOhB,sRHOPhB)
39:        sT1=sABS(sPROD(sQUOT(sT1,sDIFF(sONE,sT1))
40:       *     ,sSUM(sGFO,sHFO)))
41:        vhBP(NERR)=error(rMAXABS(sSUM(sCONST(rG(vhB)),sT1)),rH(vhB))
42:
43:        RETURN
44:        END
```

The structure of this subroutine is trivial. In L19 we calculate $-F(\Gamma) + F(\Gamma_p)$ and sum the result in L25-26 to $\Delta J = J - J_{P,thin}$. In L32 we bound ΔF using the subroutine FBNDPREIM. All these bounds are established on the domain $D(-x_{oh_B}, \rho_{h_B}^+)$ and are bounds on the supremum of the corresponding functions. We use again the inequality (5.7.34) to bound ℓ^1-norm on $D(-x_{oh_B}, \rho_{h_B})$ (L38-41). This completes the discussion of the subroutine FADDGENERAL itself and we discuss now the subroutine FBNDIM in detail, which estimates ΔJ_F.

```
 1:       SUBROUTINE FBNDIM(sC,sR,sGGAMMA,sHGAMMA,sZ,sETA,sGJFO,sHJFO)
 2:       INCLUDE BIGN.STUFF
 3:       COMMON/DOMAIN/sRHOGO,sRHOGAMMA,sXOh,sCENTh,sRHOh,sCENTG,sRHOG
 4:      *             ,sRHOhB,sRHOPhB,sXOhB,sPHIO,sxKAPPA,sKAPPA
 5:      *             ,sRHOP,sRHO1
 6:
 7:
 8:C  Estimate abs(vFO(vGAMMAP+sGGAMMA+sHGAMMA)-vFO(vGAMMAP))
 9:C  ========================================================
10:
11:C  Calculate common factor
12:C  -----------------------
13:       sT1=sQUOT(sSUM(sHGAMMA,sGGAMMA),sRHOP)
14:       sT2=sSUM(sTWO,sQUOT(sT1,sETA))
15:       sT3=sDIFF(sONE,sQUOT(sT1,sETA))
16:       IF(DIMAG(sT3).LT.rZERO)WRITE(6,*)´error: sT3<0´,sT3
17:       sT3=sPOWER(sT3,2)
18:       sT4=sPROD(sPOWER(sINV(sETA),3),sQUOT(sT2,sT3))
19:       sT5=sPROD(sPROD(sC,sRHOP),sT4)
20:       sT6=sDIFF(sINV(sZ),sSUM(sPROD(sT4,sT1)
21:      *     ,sQUOT(sPOWER(sRHOP,2),sR)))
22:       IF(DIMAG(sT6).LT.rZERO)WRITE(6,*)´error: sT6<0´,sT6
23:       sT7=sQUOT(sT5,sPOWER(sT6,2))
24:
25:C  Calculate general
26:C  -----------------
27:       sGJFO=sUPPER(sPROD(sT7,sGGAMMA))
28:
29:C  Calculate higher order
30:C  ----------------------
31:       sHJFO=sUPPER(sPROD(sT7,sHGAMMA))
32:
33:
34:       RETURN
35:       END
```

We need to establish bounds on the difference

$\Delta F = \{F(\Gamma) - F(\Gamma_p)\}$, and since we have already the bounds ΔJ_G

and ΔJ_H which are $O(z^4)$ and $O(z^{N-3})$ respectively, we also

split ΔJ_F into $\Delta J_{F,G} + \Delta J_{F,H}$ and require $\Delta J_{F,G}(z) = O(z^4)$,

$\Delta J_{F,H}(z) = O(z^{N-3})$. We then want to bound (L27)

$$\sup_{z \in \delta} \left| \frac{\Delta J_{F,G}}{(z/\rho^+)^4} \right| \equiv \text{sGJF}\emptyset \ , \tag{5.7.43}$$

and (L31)

$$\sup_{z \in \delta} \left| \frac{\Delta J_{F,H}}{(z/\rho^+)^{N-3}} \right| \equiv \text{sHJF}\emptyset \ . \tag{5.7.44}$$

This we achieve as follows. Writing $z = |z|e^{i\varphi}$ we get

$$\left| (F(\Gamma_p + \Delta\Gamma) - F(\Gamma_p))(z) \right|$$

$$= \left| \int_0^\infty e^{-t/\Gamma_p^2(z)} (e^{-\eta(z)t} - 1) f_{B,\mathcal{L}}(t) dt \right|$$

$$= \left| z^2 \int_0^{e^{2i\varphi} \cdot \infty} e^{-t(z/\Gamma_p(z))^2} \eta(z) z^2 t \cdot \left(\frac{e^{-\eta(z)z^2 t} - 1}{\eta(z) z^2 t} \right) f_{B,\mathcal{L}}(z^2 t) dt \right|$$

$$\leq |z|^2 \int_0^{e^{2i\varphi} \cdot \infty} e^{-\text{Re}(t(z/\Gamma_p(z))^2)} \cdot |\eta(z) \cdot z^2 t| \cdot$$

$$\cdot \left| \frac{e^{-\eta(z)z^2 t} - 1}{\eta(z) z^2 t} \right| \cdot |f_{B,\mathcal{L}}(z^2 t)| \, |dt| \ , \tag{5.7.45}$$

where $\eta(z) = (\frac{1}{\Gamma(z)})^2 - (\frac{1}{\Gamma_p(z)})^2$. It is useful to define

$H_p(z) = \frac{1}{z}\Gamma_p(z)$. We shall need an upper bound on $\sup_{z \in \delta} |(H_p(z))^{-2}|$

and we get this by observing that

$$\sup_{z \in \delta} | (H_P(z))^{-2} | \leq | (\frac{1}{H^u(\rho^+ z/\rho_\Gamma)})^2 |_1 =: (\frac{1}{\eta})^2 , \qquad (5.7.46)$$

where the inequality is true since all terms in $\Delta\Gamma$ are on the computer always counted with the most pessimistic sign and thus

$$| (\frac{1}{H_P^u(\rho^+ z/\rho_\Gamma)})^2 |_1 \leq | (\frac{1}{H^u(\rho^+ z/\rho_\Gamma)})^2 |_1 .$$

We use $\eta \equiv$ sETA. See (5.7.8). We next bound the exponential in (5.7.45), i.e. we establish a lower bound for $\text{Re}(t(z/\Gamma_p(z))^2)$. Denoting $t = (a + ib)|z|$ and $(1/H_p(z))^2 = c + id$ we see that

$$\text{Re}(\frac{t}{H_p(z)})^2 \geq ac - |bd| \geq \frac{|t|}{r} , \qquad (5.7.47)$$

where $r \equiv$ sZ has already been given in (5.7.7) and is obtained as follows :

$$\inf_{z \in \delta} (\text{Re}((\frac{1}{H_p(z)})^2)) \geq 1 - \sup_{z \in \delta} | (\frac{1}{H_p(z)})^2 - 1 |$$

$$\geq 1 - | (\frac{1}{H^u(\rho^+ z/\rho_\Gamma)})^2 - 1 |_B =: \hat{c}$$

and similarly,

$$\sup_{z \in \delta} |\text{Im}((\frac{1}{H_p(z)})^2)| \leq | (\frac{1}{H^u(\rho^+ z/\rho_\Gamma)})^2 - 1 |_B =: \hat{d} .$$

If $z \in \delta$ then $a \geq \cos 2\varphi_o$ and $b \leq \sin 2\varphi_o$. Therefore we find for the bound of $\frac{1}{r}$ in (5.7.47)

$$\frac{1}{r} = (\cos 2\varphi_o)\hat{c} - (\sin 2\varphi_o)\hat{d} \ .$$

We next bound the function $\eta(z)$ in (5.7.45). We have

$$|\eta(z)| = |((\Gamma_p + \Delta\Gamma)^{-2} - (\Gamma_p)^{-2})(z)|$$

$$= |((\frac{1}{\Gamma_p})^3 \frac{2+\Delta\Gamma/\Gamma_p}{(1+\Delta\Gamma/\Gamma_p)^2}\Delta\Gamma)(z)|$$

$$\leq (\frac{1}{\eta})^3 (\frac{2+\frac{1}{\eta}X}{(1-\frac{1}{\eta}X)^2})(\frac{1}{\rho^+})^3 \cdot |\frac{\Delta\Gamma(z)}{(2/\rho^+)^3}| \ , \qquad (5.7.48)$$

where $X := \sup\limits_{z\in\delta}|\Delta\Gamma/z|$, and η is defined in (5.7.46).

Define now $Y := \sup\limits_{z\in\delta}|\Delta\Gamma(z)/(z/\rho^+)^3|$, then from (5.7.10) and (5.7.11) it follows that

$$Y = \rho_\Gamma(\frac{\rho^+}{\rho_\Gamma})^3|\Delta\Gamma_G^u|_1$$

$$+ \rho_\Gamma(\frac{\rho^+}{\rho_\Gamma})^{N-3}|\Delta\Gamma_H^u|_1 \equiv \text{SGGAMMA} + \text{SHGAMMA} \ . \qquad (5.7.49)$$

Thus we get

$$\sup\limits_{z\in\delta}|\eta(z)| \leq (\frac{1}{\eta})^3 \frac{2+\frac{1}{\eta}X}{(1-\frac{1}{\eta}X)^2}(\frac{1}{\rho^+})^3 Y =: ZY \ . \qquad (5.7.50)$$

Similarly, since $|(\exp(z) - 1)/z| \leq \exp(|z|)$, we see that

$$\sup\limits_{z\in\delta}|\frac{e^{-\eta(z)z^2t}-1}{\eta(z)z^2t}| \leq \exp(ZY\cdot(\rho^+)^2\cdot|t|) \ . \qquad (5.7.51)$$

Finally we have that $|f_{B,\mathcal{L}}(t)| \leq C_{\mathcal{L}} \cdot \exp(|t|/R_{\mathcal{L}})$ and we are thus led to the following bound on (5.7.45):

$$(F(\Gamma_p + \Delta\Gamma) - F(\Gamma_p))(z)|$$

$$\leq |z|^4 \int_0^\infty e^{-t/r} tz \left| \frac{\Delta\Gamma(z)}{(z/\rho^+)^3} \right| e^{ZY(\rho^+)^2 t} C_{\mathcal{L}} e^{t\rho^{+2}/R_{\mathcal{L}}} dt$$

$$= |z|^4 \cdot \left| \frac{\Delta\Gamma(z)}{(z/\rho^+)^3} \right| \cdot ZC_{\mathcal{L}} \left(\frac{1}{\frac{1}{r} - ZY(\rho^+)^2 - (\rho^+)^2/R_{\mathcal{L}}} \right)^2 . \tag{5.7.52}$$

If we consider $F(\Gamma_p + \Delta\Gamma) - F(\Gamma_p)$ and rewrite it as in (5.7.52) but without absolute values, we see that there is an explicit factor $\Delta\Gamma(z)$. Therefore, splitting $\Delta\Gamma = \Delta\Gamma_G + \Delta\Gamma_H$ leads naturally to a decomposition $\Delta J_F = \Delta J_{F,G} + \Delta J_{F,H}$. We have the bounds

$$|\Delta J_{F,G}| \leq |z|^4 \cdot \left| \frac{\Delta\Gamma_G}{(z/\rho^+)^3} \right| \cdot Z \cdot C_{\mathcal{L}} \cdot \left(\frac{1}{\frac{1}{r} - ZY(\rho^+)^2 - (\rho^+)^2/R_{\mathcal{L}}} \right)^2 ,$$

$$|\Delta J_{F,H}| \leq |z|^4 \cdot \left| \left(\frac{z}{\rho^+} \right)^{N-2} \right| \cdot \left| \frac{\Delta\Gamma_H}{(z/\rho^+)^{N+1}} \right| \cdot Z \cdot C_{\mathcal{L}} \cdot \left(\frac{1}{\frac{1}{r} - ZY(\rho^+)^2 - (\rho^+)^2/R_{\mathcal{L}}} \right)^2 ,$$

and therefore, and using again (5.7.10) and (5.7.11) and the definitions (5.7.43) and (5.7.44),

$$\sup_{z \in \delta} \left| \frac{\Delta J_{F,G}}{(z/\rho^+)^4} \right| \leq \rho^+ \left(\frac{1}{3} \frac{2 + X/\eta}{(1-X/\eta)^2} \right) \cdot \left\{ \rho_\Gamma \left(\frac{\rho^+}{\rho_\Gamma} \right)^3 |\Delta\Gamma_G^u|_1 \right\}$$

$$\cdot C \cdot \left(\frac{1}{r} - YZ(\rho^+)^2 - \frac{(\rho^+)^2}{R_{\mathcal{L}}} \right)^{-2} , \tag{5.7.53}$$

$$\sup_{z \in \delta} \left| \frac{\Delta J_{F,G}}{(z/\rho^+)^4} \right| \leq \rho^+ \left(\frac{1}{\eta^3} \frac{2+X/\eta}{(1-X/\eta)^2} \right) \cdot \{ \rho_\Gamma \left(\frac{\rho^+}{\rho_\Gamma} \right)^{N+1} |\Delta\Gamma_H^u|_1 \}$$

$$\cdot C \cdot \left(\frac{1}{r} - YZ(\rho^+)^2 - ('\rho^+)^2/R_{\mathcal{L}} \right)^{-2} . \qquad (5.7.54)$$

Note furthermore that $X = \sup_{z \in \delta} |\Delta\Gamma(z)/z| = Y/\rho^+$. We refer now to the lines of the subroutine FBNDIM. We have $X \equiv$ sT1 (L13), $(\frac{1}{\eta^3}(2 + X/\eta)/(1 - X/\eta)^2) \equiv$ sT4 (L18). In L19 we multiply with ρ^+ and C. Note that $YZ(\rho^+)^2 \equiv$ sT4 $*_s$ sT1. This explaines L20-21. The remaining lines of the subroutine are selfexplaining. Again we test (L16, L22) that all the denominators which occur are positive. This completes our discussion of the subroutine FBNDIM. We discuss now the subroutine FBNDPREIM, which solves Ecalle's equation, given bounds on the inhomogeneity. This will complete the discussion of FADDGENERAL.

```
 1:        SUBROUTINE FBNDPREIM(sGJ,sHJ,sRHOP,sL5,sL6,sGFO,sHFO)
 2:        INCLUDE BIGN.STUFF
 3:
 4:
 5:C  Invert Ecalle's equation
 6:C  ========================
 7:
 8:C  Calculate sup over sector; contribution from general
 9:C  ----------------------------------------------------
10:        sGFO=sPROD(sGJ,sSUM(sONE,sQUOT(sPOWER(sINV(sRHOP),2)
11:     *      ,sPROD(sL5,sL6))))
12:
13:C  Calculate sup over sector; contribution from higher order
14:C  ---------------------------------------------------------
15:        sHFO=sPROD(sHJ,sSUM(sONE,sQUOT(sPOWER(sINV(sRHOP),2)
16:     *      ,sPROD(sQUOT(siCONST(N-5),sTWO),sPROD(sL5
17:     *      ,sPOWER(sSQRT(sL6),N-5)))))))
18:
19:
20:        RETURN
21:        END
```

This subroutine deals with two functions ΔJ_G and ΔJ_H analytic on δ, $\Delta J_G(z) = 0(z^4)$ and $\Delta J_H = 0(z^{N-3})$. The input consists of

$$\sup_{z \in \Delta} \left| \frac{\Delta J_G(z)}{(z/\rho^+)^3} \right| \equiv sGJ \ , \tag{5.7.55}$$

$$\sup_{z \in \Delta} \left| \frac{\Delta J_H(z)}{(z/\rho^+)^{N-3}} \right| \equiv sHJ \ . \tag{5.7.56}$$

The task is to find bounds on functions ΔF_G , ΔF_H , which satisfy $(\Delta F_G + \Delta F_H) \circ \Gamma - (\Delta F_G + \Delta F_H) = \Delta J_G + \Delta J_H$, on Δ. ΔF_G and ΔF_H will turn out to be $O(z^2)$ and $O(z^{N-5})$ respectively and we call

$$\sup_{z \in \Delta} \left| \frac{\Delta F_G(z)}{(z/\rho^+)^2} \right| \equiv sGF\emptyset \ , \tag{5.7.57}$$

and

$$\sup_{z \in \Delta} \left| \frac{\Delta F_H(z)}{(\rho^+/z)^{N-5}} \right| \equiv sHF\emptyset \ , \tag{5.7.58}$$

respectively. Remember that we have $\inf_{z \in \Delta} |\mathrm{Re}(R(z))| \equiv sL5$ (see (5.7.6)), where $R(z) = (1/\Gamma(z))^2 - (1/z)^2$ (see (5.7.4)). We use now that $\Gamma(\Delta) \subset \Delta$ and get

$$|\Gamma^n(z)| \leq |n \cdot L_5 + L_6/|z|^2)^{-1/2}$$

for all $z \in \Delta$. (Remember that $L_6 = \cos 2\varphi_0 \equiv sL6$.) We now intend to write

$$F_G^{}(z) = - \sum_{n=0}^{\infty} \Delta \dot{J}_G(\Gamma^n(z)) \ ,$$

and to just bound the sup of this sum. We clearly have

$$\sum_{n=0}^{\infty} |\Gamma^n(z)|^m \leq |z|^m + \sum_{n=1}^{\infty} (n \cdot L_5 + L_6/|z|^2)^{-m/2}$$

$$\leq |z|^m + \int_0^{\infty} (xL_5 + L_6/|z|^2)^{-m/2} dx$$

$$= |z|^m + \frac{1}{L_5(m/2-1)} (\frac{1}{L_6})^{m/2-1} |z|^{m-2} \quad .$$

Since $\Delta J_G = O(z^4)$ and $\Delta J_H = O(z^{N-3})$ we get the bounds

$$\sup_{z \in \delta} |\frac{\Delta F_G}{(z/\rho^+)^2}| \leq \sup_{z \in \delta} |\frac{\Delta J_G}{(z/\rho^+)^4}| \cdot \sup_{z \in \delta} |\frac{1}{(z/\rho^+)^2}| \cdot \sum_{n=0}^{\infty} (\frac{1}{\rho^+})^4 |\Gamma^n(z)|^4$$

$$\leq \sup_{z \in \delta} |\frac{\Delta J_G}{(z/\rho^+)^4}| \cdot (\frac{1}{\rho^+})^4$$

$$\cdot \sup_{z \in \delta} |\frac{1}{(z/\rho^+)^2}| (|z|^4 + \frac{1}{L_5 L_6} |z|^2)$$

$$= \sup_{z \in \delta} |\frac{\Delta J_G}{(z/\rho^+)^4}| (1 + \frac{(1/\rho^+)^2}{L_5 L_6}) , \qquad (5.7.59)$$

and similarly

$$\sup_{z \in \delta} |\frac{\Delta F_H}{(z/\rho^+)^{N-3}}| \leq \sup_{z \in \delta} |\frac{\Delta J_H}{(z/\rho^+)^{N-3}}| (1 + \frac{(1/\rho^+)^2}{L_5(N-5)/2} (\frac{1}{L_6})^{(N-5)/2})$$

$$(5.7.60)$$

This are exactly the bounds which are implemented in L10-11 and L15-17 of FBNDPREIM. This completes our discussion of the subroutine FBNDPREIM and hence of the subroutine FADDGENERAL. In addition this completes our discussion of Borel summation with rigorous bounds. The vector $vhB \equiv h_B^u$ is now completed up to those terms coming from the singular part F_{sing}, which has been split off in the subroutine FREDUCE above. These terms are now added in the subroutine FCOMPLETE which we discuss next.

5.7.6 Addition of the singular terms

```
 1:        SUBROUTINE FCOMPLETE(vhBH,sA0,sC0,sD0,vhB)
 2:        INCLUDE BIGN.STUFF
 3:        COMMON/DOMAIN/sRHOG0,sRHOGAMMA,sX0h,sCENTh,sRHOh,sCENTG,sRHOG
 4:       *            ,sRHOhB,sRHOPhB,sX0hB,sPHI0,sxKAPPA,sKAPPA
 5:       *            ,sRHOP,sRHO1
 6:        DIMENSION vhBH(1),vhB(1)
 7:        DIMENSION vT1(NDEG)
 8:
 9:
10:C  Store Borel part of solution in result
11:C  ======================================
12:
13:        CALL FEQUAL(vhBH,vhB)
14:
15:
16:C  Add irregular part of solution
17:C  ==============================
18:
19:C  Add term with sA0
20:C  -----------------
21:        CALL FSETZERO(vT1)
22:        vT1(1)=sONE
23:        vT1(2)=sNEG(sQUOT(sPROD(sTWO,sRHOhB),sX0hB))
24:        vT1(3)=sPROD(sQUOT(sRHOhB,sX0hB),sQUOT(sRHOhB,sX0hB))
25:        CALL FINVERS(vT1,vT1)
26:        vT1(1)=sZERO
27:        CALL FsMULADD(vT1,sQUOT(sA0,sPROD(sX0hB,sX0hB)),vhB)
28:
29:C  Add terms with sC0
30:C  ------------------
31:        CALL FSETLOG(vT1,sNEG(sQUOT(sRHOhB,sX0hB)))
32:        CALL FsMULADD(vT1,sPROD(sTWO,sC0),vhB)
33:
34:C  Add term with sD0
35:C  -----------------
36:        vhB(2)=sSUM(vhB(2),sPROD(sD0,sRHOhB))
37:
38:
39:        RETURN
40:        END
```

This subroutine is completely trivial. In L21-27 we construct

the function a_o/z^2 on the domain $D(-x_{oh_B}, \rho_{h_B})$ and add it to

h_B^u. Similarly in L31-32 for $c_o logz^2$ and in L36 for $d_o \cdot z$.

This completes our discussion of this subroutine.

We close this section by noting that we have explained all

lines of the subroutine FvKI (for the case IFLAG = 0). The

operator K is now defined and $\varepsilon = \|K(\{A,G\}) - \{A,G\}\|$ can be

calculated. In the next section we show how the tangent map DK

is implemented.

5.8. The implementation of the tangent map DK

Below, we use the notation L_M to refer to line numbers of

the main program. (See Section 5.5 for the listing of the main

program.) Let $\{A,G_o\} \in U_\beta \subset A_1$ as described in Theorem 5.4.3 of

Section 5.4 and where $\beta \equiv$ rBETA (L_M154-155). We want to establish

a bound on the operator norm of $DK_{\{A,G_o\}}$, the tangent map of K

at $\{A,G_o\}$. This bound should be uniform with respect to

$\{A,G_o\} \in U_\beta$. We have already explained in Section 5.5 that this

problem can be reduced to calculate on the computer

$|DK_{\{A,G_o\}}(\{\delta A, \delta G_o\})|_1$ for a vector $vK\emptyset \equiv \{A,G_o\}$ (L_M75-81, L_M161)

which represents <u>any</u> element $\{A,G_o\} \in U_\beta$ and where furthermore the

basis elements of the tangent space A, $\{\delta A, \delta G_o\} = \{\delta A, \delta G_o\}_i$,

$i = 1,...,\infty$ are represented by a <u>finite</u> number of vectors. The

equations defining DK have been given in Section 5.3. It is easy

to see that many of the subexpressions occurring in these equa-

tions do only depend on $\{A,G_o\} \in U_\beta$ (and therefore on just one

vector, namely $vK\emptyset$ as defined above) and not on the element

$\{\delta A, \delta G_0\}$ in the tangent space. For example all the vectors which are furnished by the subroutine FvKI occur as subexpressions of these equations.

In order to save computer time we therefore proceed on the computer as follows : We first calculate $(L_M 194-523)$ in a preparatory step all those vectors which are independent of the tangent vector chosen and "store" the results in the common-block COMMON/STRES1/. In a second step $(L_M 525-588)$ we effectively apply the tangent map DK by calling the subroutine FvDKI.

Note again that these calculations are not all carried out with the same degree N for the polynomials as we have already mentioned in Section 5.5. The corresponding degrees are chosen in $L_M 180$. Furthermore in $L_M 182-188$ we adapt the vector vKØ to these degrees. The bound on the norm of the tangent map DK is called $\rho \equiv$ sRHO $(L_M 167, L_M 556, L_M 588)$. In $L_M 600$ finally we check if the inequality $\varepsilon < (1 - \rho)\beta$ is satisfied, which decides if our program does or does not prove the existence of a fixed print for K in U_β.

After these preliminary remarks we discuss now all these "preparatory calculations" which are carried out by calling the subroutine FvKI with IFLAG = 1 $(L_M 194)$; (see Section 5.6 for the listing of the subroutine FvKI) : We only need to discuss all those parts which are changed if one calls the subroutine with IFLAG = 1 instead of IFLAG = 0. A trivial change is that we produce output for information on bounds for A $\equiv$ sAAA (L22),

$\kappa \equiv$ sKAPPA (L103), $a_o \equiv$ sAØ (L262), $c_o \equiv$ sCØ (L263), $d_o \equiv$ sDØ (L264),

$\tau \equiv$ sTAU (L321) and $x_o \equiv$ sXØ (L334-338). Furthermore we do not

recalculate the functions $M_{1,k}$ and $M_{2,k}$ since they are inde-

pendent of the vector vKØ (L271-282). A somewhat less trivial

change comes from the fact that we are also calling the subroutine

CREDUCE with IFLAG = 1 (L259-261). A quick check of the listing

of CREDUCE (see Section 5.7.3) reveals however that nothing

changes there but that in turn the subroutine FREDUCE gets called

with IFLAG = 1, and we discuss those changes next. (See Section

5.7.3 for the listing of the subroutine FREDUCE.) In fact, we

calculate in addition to the function $J(z)$ its derivative $J'(z)$.

There are however two points which need special care : First of

all, we have to restrict analyticity domains when we are taking

derivatives since we want to work with functions with finite ℓ_1-

norm. The function J is analytic on the disk $D(0, \rho_\Gamma)$ and it is

related to its normalized form J^u by the equation $J(z) = J^u(z/\rho_\Gamma)$.

We choose to restrict the analyticity domain for the derivative

to the disk $D(0, \rho^+)$. We get therefore for the normalized form

of the function J' (L49-59) :

$$(J')^u(z) = J'(\rho^+ z) = \frac{1}{\rho_\Gamma}(J^u)'(\frac{\rho^+}{\rho_\Gamma}z) \equiv vDZJ . \tag{5.8.1}$$

(We calculate in L49-59 in addition some ℓ^1-bounds which we use

below.) Remember now that we have used for the function J the

decomposition $J = J_{P,thin} + \Delta J_G + \Delta J_H$, with $\Delta J_G(z) = O(z^4)$

and $\Delta J_H(z) = O(z^{N-2})$. This decomposition implies a similar de-

composition of J', namely $J'(z) = (J_{P,thin})'(z) + (\Delta J_G)'(z) +$

$+ (\Delta J_H)'(z)$, with $(\Delta J_G)'(z) = O(z^3)$ and $(\Delta J_H)'(z) = O(z^{N-3})$.

(It is now clear, why we have decided to use $O(z^{N-3})$ and not $O(z^{N-2})$ for ΔJ in Section 5.7.4. Namely we need also to bound derivatives of functions which are $O(z^{N-2})$.) Unfortunately this decomposition of J' is not sufficient for our purposes below, and we shall decompose J' in a different way. We write

$J'(z) = (J')_P(z) + \Delta(J')(z)$, and we define $(J')_P(z) = (J_{P,thin})'(z)$

and

$$\Delta(J')(z) = J_1^o z^3 + J_2^o z^4 + J_3^o z^5 +$$

$$+ \Delta(J')_G(z) + \Delta(J')_H(z) , \qquad (5.8.2)$$

where $\Delta(J')(z) = (\Delta J_G + \Delta J_H)'(z)$, $\Delta(J')_H(z) = (\Delta J_H)'(z)$ and $\Delta(J')_G(z) = O(z^6)$. In order to unburden the notation we shall write J_P', $\Delta J'$, $\Delta J_G'$ and $\Delta J_H'$ instead of $(J')_P$, $\Delta(J')$, $\Delta(J')_G$ and $\Delta(J')_H$. The reader should distinguish this from expressions like $(\Delta J_G)'$, $(\Delta J_H)'$, etc. We have (L75-77)

$$(J_P')^u(z) = \frac{1}{\rho_\Gamma}(J_{P,thin}^u)'(\frac{\rho^+}{\rho_\Gamma}z) \equiv vxDZJ \qquad (5.8.3)$$

and (L117-122)

$$\sup_{z \in \delta} \left| \frac{\Delta J_H'(z)}{(z/\rho^+)^{N-3}} \right| \leq |(\Delta J_H')^u(.)|_B \equiv sHDZJ \qquad (5.8.4)$$

and (L135-138)

$$\sup_{z \in \delta} \left| \frac{\Delta J_H'(z)}{(z/\rho^+)^6} \right| \leq |(\Delta J_G')^u(.)|_B \equiv sGDZJ \qquad (5.8.5)$$

and furthermore we calculate bounds on the coefficients

$$J_1^o \equiv sJ00 , \qquad\qquad (5.8.6)$$

$$J_2^o \equiv sJ01 , \qquad\qquad (5.8.7)$$

and

$$J_3^o \equiv sJ02 , \qquad\qquad (5.8.8)$$

in L132, L133 and L134 respectively.

This completes our discussion of these preparatory calculations which are carried out by calling the subroutine FvKI with IFLAG = 1. We discuss now right away the subroutine FvDKI. Those preparatory calculations which are carried out in the main program will be explained on their first occurrence. Again we start the discussion by giving first the listing of the program.

```
1:        SUBROUTINE FvDKI(vDKO,vDKOI)
2:        INCLUDE BIGN.STUFF
3:        COMMON/DOMAIN/sRHOGO,sRHOGAMMA,sXOh,sCENTh,sRHOh,sCENTG,sRHOG
4:       *             ,sRHOhB,sRHOPhB,sXOhB,sPHIO,sxKAPPA,sKAPPA
5:       *             ,sRHOP,sRHO1
6:        COMMON/SCALE /sXOO
7:        COMMON/SWITCH/ITEST
8:        INCLUDE BIGN.STRES1
9:        DIMENSION vDKO(1),vDKOI(1)
10:        DIMENSION vDGO(NDEG),vDGAMMA(NDEG),vDRHS(NDEG),vDhB(NDEG)
11:       *          ,vDh(NDEG),vDG(NDEG),vDGMOEBIUS(NDEG)
12:       *          ,vDGOI1(NDEG),vDGOI2(NDEG),vDGOI(NDEG)
13:        DIMENSION xDFO(NDEG),xDJ1(NDEG),xDJ(NDEG)
14:        DIMENSION vT1(NDEG),vT2(NDEG)
15:        DIMENSION xT1(NDEG)
16:
17:
18:C  UNPACK isomorphism
19:C  ==================
20:
21:        CALL UNPACK(vDKO,sDAAA,vDGO)
22:        vDGO(2)=sZERO
23:
24:
25:C  Ecalle's equation for vDh
26:C  =========================
27:
28:C  Produce function vDGAMMA and normalize it
29:C  -----------------------------------------
30:        CALL FsDILAT(vDGO,sQUOT(sRHOGAMMA,sRHOGO),vT1)
31:        CALL FMULT(vDZGOGO,vT1,vT1)
32:        CALL FCOMP(vDGO,vGORED,vT2)
33:        CALL FADD(vT2,vT1)
34:        CALL FsMULT(vT1,sQUOT(sRHOGO,sRHOGAMMA),vDGAMMA)
35:        vDGAMMA(1)=sZERO
36:        vDGAMMA(2)=sZERO
37:        vDGAMMA(3)=sZERO
38:        vDGAMMA(5)=sZERO
39:
40:C  Calculate function vDRHS
41:C  ------------------------
42:        CALL FSHFTL(vDGAMMA,vT1)
43:        vT1(1)=sZERO
44:        CALL FSHFTL(vT1,vT1)
45:        vT1(1)=sZERO
46:        CALL FSHFTL(vT1,vT1)
47:        CALL FMULT(vACD,vT1,vDRHS)
48:
49:C  Calculate power series xDJ1 and Borel bounds
50:C  --------------------------------------------
51:C..calculate power series
52:        CALL CvTOxe(vDGAMMA,xT1,eT1)
53:        CALL CSHFTL(xT1,xT1)
54:        ITEST=0
55:        CALL FsDILAT(xT1,sINV(sRHOGAMMA),xT1)
56:        ITEST=1
57:        CALL CMULT(xDZFO,xT1,xDJ1)
58:C..calculate bound sGDJ1 and sHDJ1
59:        CALL FsDILAT(vDGAMMA,sRHO1,vT1)
60:        sT1=sPROD(sRHOGAMMA,sCONST(rL1BOUND(vT1)))
61:        sGDJ1=sPROD(sGDZDFO,sT1)
62:        sHDJ1=sPROD(sHDZDFO,sT1)
63:
64:C  Calculate regular inhomogeneous part
65:C  ------------------------------------
```

```
 66:          CALL CDREDUCE(xDJ1,N,sGDJ1,sHDJ1,vDRHS,sDA0,sDC0,sDD0,xDJ
 67:      *                ,sGDJ,sHDJ
 68:      *                ,vDGAMMA,sCDF0,sEDF0)
 69:
 70:C Produce power series solution of Ecalle´s equation
 71:C -------------------------------------------------
 72:          CALL CxF0(xGAMMA,xDJ,xDF0,N)
 73:
 74:C Calculate Loeffel/Borel-bounds
 75:C -----------------------------
 76:          sDC=sUPPER(sPROD(sCDF0,sQUOT(sFOUR,sxKAPPA)))
 77:          sDR=sLOWER(sINV(sPROD(sEDF0,sQUOT(sFOUR,sxKAPPA))))
 78:
 79:C Sum the formal power series part of DF0 to the sectorial solution
 80:C ----------------------------------------------------------------
 81:          CALL FSUMBOREL(xDF0,N,sDC,sDR,vDhB)
 82:
 83:C Add contribution of general term
 84:C -------------------------------
 85:          CALL FADDGENERAL(vDhB,sGDJ,sHDJ,sDC,sDR,sGGAMMA,sHGAMMA
 86:      *                    ,sZ,sETA,sL5,sL6,vDhB)
 87:
 88:C Restore irregular inhomogeneous part
 89:C -----------------------------------
 90:          IF( rMAXABS(sDA0).GT.rZERO
 91:         *.OR.rMAXABS(sDC0).GT.rZERO
 92:         *.OR.rMAXABS(sDD0).GT.rZERO)
 93:         *CALL FCOMPLETE(vDhB,sDA0,sDC0,sDD0,vDhB)
 94:
 95:C Analytic bootstrap
 96:C -----------------
 97:          CALL FCOMP(vDG0,v0ARG,vT1)
 98:          DO 1 I=1,IBOOT
 99:          CALL FMULT(vT1,vDZG0ARG(1,I),vT1)
100:          CALL FCOMP(vDG0,vARG(1,I),vT2)
101:          CALL FADD(vT2,vT1)
102:        1 CONTINUE
103:          CALL FMULT(vT1,vDZhLASTG0,vDh)
104:          CALL FCOMP(vDhB,vLASTG0,vT1)
105:          CALL FADD(vT1,vDh)
106:          vDh(1)=sZERO
107:
108:
109:C Add variation due to sDAAA
110:C -------------------------
111:          IF(rMAXABS(sDAAA).GT.rZERO)CALL FsMULADD(vDAh,sDAAA,vDh)
112:
113:
114:C determine sDTAU
115:C ==============
116:
117:          sARG=sQUOT(sDIFF(sPROD(sTAU,sX00),sCENTh),sRHOh)
118:          sDTAU=sQUOT(sNEG(sPROD(sPROD(sTAU,sLOG(sTAU))),sDVALUE(1,vDh
119:      *       ,sARG))),sSUM(sPROD(sSUM(sONE,sLOG(sTAU))),sDVALUE(1,vh
120:      *       ,sARG)),sPROD(sPROD(sPROD(sTAU,sLOG(sTAU)),sQUOT(sX00
121:      *       ,sRHOh)),sDVALUE(2,vh,sARG))))
122:
123:
124:C Determine sDALPHA
125:C ================
126:
127:          sARG=sQUOT(sDIFF(sPROD(sTAU,sX00),sCENTh),sRHOh)
128:          sDALPHA=sPROD(sALPHA,sSUM(sPROD(sSUM(sQUOT(sVALUE(vh,sARG),sTAU)
129:      *          ,sPROD(sLOG(sTAU),sPROD(sQUOT(sX00,sRHOh),sDVALUE(1,vh
130:      *          ,sARG)))),sDTAU),sPROD(sLOG(sTAU),sVALUE(vDh,sARG))))
```

```
131:
132:
133:C    Calculate image vector vDG
134:C    ==========================
135:
136:        CALL FsMULT(vDTAUG,sDTAU,vT1)
137:        CALL FCOMP(vDh,vTRANS,vT2)
138:        CALL FsMULADD(vT2,sNEG(sLOG(sTAU)),vT1)
139:        vT1(1)=sSUM(vT1(1),sQUOT(sDALPHA,sALPHA))
140:        CALL FMULT(vGPCT,vT1,vDG)
141:
142:
143:C    Calculate sDAAAI
144:C    ================
145:
146:        sT1=sQUOT(sTAU,sRHOh)
147:        sT2=sPROD(sT1,sT1)
148:        sDAAAI=sPROD(sQUOT(sX00,sFOUR),sSUM(sPROD(sSUM(sPROD
149:     *        (sDVALUE(2,vh,sARG),sPROD(sT1,sQUOT(sSUM(sONE,sPROD(sTWO
150:     *        ,sLOG(sTAU))),sRHOh))),sPROD(sDVALUE(3,vh,sARG),sPROD
151:     *        (sLOG(sTAU),sPROD(sT2,sQUOT(sX00,sRHOh)))))),sDTAU)
152:     *        ,sPROD(sDVALUE(2,vDh,sARG),sPROD(sLOG(sTAU),sT2))))
153:
154:
155:C    Calculate vDGOI
156:C    ===============
157:
158:C    Substitute Moebius map in vDG
159:C    -----------------------------
160:        CALL FCOMP(vDG,vMOEBIUS,vDGMOEBIUS)
161:
162:C    Calculate vDGOI1
163:C    ----------------
164:        CALL FsMULT(vDGMOEBIUS,sQUOT(sRHOG,sRHOGO),vT1)
165:        CALL FsMULADD(vDAIGOI,sNEG(sDAAAI),vT1)
166:        CALL FEQUAL(vT1,vDGOI1)
167:
168:C    Calculate vDGOI2
169:C    ----------------
170:        CALL FsMULT(vDGMOEBIUS,sPROD(sAAAI,sRHOG),vT1)
171:        CALL FsMULADD(vGOI1,sPROD(sDAAAI,sRHOGO),vT1)
172:        CALL FsMULADD(vDAIGOI,sNEG(sPROD(sPROD(sAAAI,sRHOGO),sDAAAI)),vT1)
173:        CALL FMULT(vGOI2,vGOI2,vT2)
174:        CALL FMULT(vT2,vT1,vDGOI2)
175:
176:C    Complete construction of vDGOI
177:C    ------------------------------
178:        CALL FMULT(vDGOI1,vGOI2,vT1)
179:        CALL FMULTADD(vGOI1,vDGOI2,vT1)
180:        CALL FEQUAL(vT1,vDGOI)
181:
182:C    Normalize vDGOI properly
183:C    ------------------------
184:        vDGOI(1)=sZERO
185:        vDGOI(2)=sZERO
186:        vDGOI(3)=sZERO
187:
188:
189:C    PACK isomorphism
190:C    ================
191:
192:        vDGOI(2)=sNEG(sONE)
193:        CALL PACK(sDAAAI,vDGOI,vDKOI)
194:
195:        RETURN
196:        END
```

This subroutine is essentially a transcription of the equations (5.3.11)-(5.3.20), defining the tangent map DK, for vectors. Given $vDK\emptyset \equiv \{\delta A, \delta G_o\} \in A$ the subroutine UNPACK (L21) creates a scalar $\delta A \equiv sAAA$ and a function $(\delta G_o^u(z) - z)$ according to (5.2.6), (5.2.7). However, for elements in A the isomorphism to ℓ^1 is not given by (5.2.6), (5.2.7) but by (5.2.8), (5.2.9) and we have to add z (L22) in order to get $\delta G_o^u(z) \equiv vDG\emptyset$. In L30-38 we construct $\delta\Gamma^u(z) \equiv vDGAMMA$, see (5.3.11), where

$$G_o^{u'}(G_o^u(\frac{\rho_\Gamma}{\rho_o}z)) \equiv vDZG\emptyset G\emptyset \qquad (5.8.9)$$

has already been calculated in the preparatory step $(L_M 207-208)$. (Note that in $L_M 202-203$ we change the degree from "sequences" to "vectors" in order to act on the vectors which are stored in the common-block COMMON/STRES1/. We change this degree back in $L_M 523$. See also the discussion of the subroutines PACK and UNPACK in Section 5.2 for these matters.)

The first (and only) nontrivial step in the subroutine FvDKI is again to solve Ecalle's equation (5.3.12). This is what is done in L40-93. This problem is now even more difficult than it was in the case of the subroutine FvKI, since we have to solve this equation with $-F_o'(\Gamma)\delta\Gamma$ as an inhomogeneity (instead of simply -2), where F_o itself is already singular since it is the solution of the just mentioned equation of Ecalle's type with -2 on its right hand side. In order to give some insight into the general strategy of attack to this problem we again present a table.

Table 5.8.1

$$\delta F_o(\Gamma) - \delta F_o = -F_o'(\Gamma)\delta\Gamma =: \delta R + \delta J_1 + \Delta\delta J_{1,G} + \Delta\delta J_{1,H}$$

where

$\delta R := -F_{sing}'(\Gamma)\delta\Gamma$ (given as a vector)

$\delta J_1 := -F'(\Gamma_p)\delta\Gamma$ (given as a power series, and bounds on

$\qquad\qquad f_{B,\mathcal{L}}$ via COMMON)

$\Delta\delta J_{1,G} + \Delta\delta J_{1,H} := -\Delta F'(\Gamma)\delta\Gamma - (F'(\Gamma)-F'(\Gamma_p))\delta\Gamma$ (given as
$\qquad\qquad\qquad\qquad\qquad\qquad\qquad\qquad\qquad\qquad$ sup on δ)

FVDKI (L42-62)

and where $\Delta F'$ is the solution of

$$\Delta F'(\Gamma)\Gamma' - \Delta F' = \Delta J_{tot}' = \Delta J' + F'(\Gamma)\Delta\Gamma' + (F'(\Gamma)-F'(\Gamma_p))\Gamma_p'$$

__Ansatz__ : $\delta F_o = \delta F_{sing} + \delta F_{reg}$, where

$\qquad\qquad \delta F_{sing} = \delta a_o/z^2 + \delta c_o \log z^2 + \delta d_o z$

__Equations__ : $\delta F_{reg}(\Gamma) - \delta F_{reg} = \delta J + \Delta\delta J_G + \Delta\delta J_H$, where

$\qquad\qquad \delta J = \delta J_1 + \delta J_{2,thin}$, where

$\qquad\qquad \delta J_{2,thin} = [\delta R - (\delta F_{sing}(\Gamma)-\delta F_{sing})]_{P,thin}$

$\qquad\qquad \Delta\delta J_G = \Delta\delta J_{1,G} + \Delta\delta J_{2,G}$, where

$\qquad\qquad \Delta\delta J_{2,G} = [\delta R - (\delta F_{sing}(\Gamma)-\delta F_{sing})]_G$

$\qquad\qquad \Delta\delta J_H = \Delta\delta J_{1,H} + \Delta\delta J_{2,H}$, where

$\qquad\qquad \Delta\delta J_{2,H} = [\delta R - (\delta F_{sing}(\Gamma)-\delta F_{sing})]_H$

FDREDUCE

__Ansatz__ : $\delta F_{reg} = \delta F + \Delta\delta F$

__Equations__ : $\delta F(\Gamma_p) - \delta F = \delta J$

$\qquad\qquad \Delta\delta F(\Gamma) - \Delta\delta F = \Delta\delta J_G + \Delta\delta J_H - (\delta F(\Gamma) - \delta F(\Gamma_p))$

__Solutions__ : $\Gamma_p, \delta J \to \hat{\Gamma}_p(x)$, $\delta\hat{J}(x)$ and bound on

$\qquad\qquad |([\delta F]_{normal\ form})_B(t)| \le C_{\delta F\emptyset}\exp(E_{\delta F\emptyset}|t|)$

CDREDUCE

__Solutions__ : i) $\delta\hat{F}(\hat{\Gamma}_p(x)) - \delta\hat{F}(x) = \delta\hat{J}(x)$ $\qquad\qquad$ CxFØ

$\qquad\qquad$ ii) $\delta\hat{F}(x)$, $C_{\delta F\emptyset}$, $E_{\delta F\emptyset} \to \delta h_B^u$ $\qquad\qquad$ FSUMBOREL

$\qquad\qquad$ iii) $\Delta\delta J_{F\emptyset,G} + \Delta\delta J_{F\emptyset,H} = -(F(\Gamma)-F(\Gamma_p))$ $\qquad$ FBNDIM

$\qquad\qquad$ iv) $\Delta\delta J_{tot,G} = \Delta\delta J_G + \Delta\delta J_{F\emptyset,G}$,

$\qquad\qquad\qquad \Delta\delta J_{tot,H} = \Delta\delta J_H + \Delta\delta J_{F\emptyset,H}$

$\qquad\qquad$ v) $\Delta\delta F(\Gamma) - \Delta\delta F = \Delta\delta J_{tot,G} + \Delta\delta J_{tot,H}$,

$\qquad\qquad\qquad \Delta\delta F = \Delta\delta F_G + \Delta\delta F_H$ $\qquad\qquad$ FBNDPREIM

$\qquad\qquad$ (vi) $\delta F_o = \delta F_{reg} + \delta F_{sing}$ $\qquad\qquad$ FCOMPLETE)

FADDGENERAL

We discuss now L42-62 of FvDKI. In L42-47 we calculate the function

$$\delta R(z) = -F'_{sing}(\Gamma(z))\delta\Gamma(z) . \qquad (5.8.10)$$

Since $F_{sing}(z) = a_o/z^2 + c_o \log z^2 + d_o z$, we have that

$$F'_{sing}(z) = -2a_o/z^3 + 2c_o/z + d_o , \qquad (5.8.11)$$

and therefore

$$-F'_{sing}(\Gamma(z)) = 2a_o/\Gamma^3(z) - 2c_o/\Gamma(z) - d_o .$$

We define the function $f_{acd}(z) = -z^3 \cdot F'_{sing}(\Gamma(z))$ and $f^u_{acd}(z) = (\frac{1}{\rho_\Gamma})^2 f_{acd}(\rho_\Gamma z)$. We get

$$f^u_{acd}(z) = (\frac{1}{\rho_\Gamma})^2 \{2a_o(\frac{1}{H^u(z)})^3 - 2c_o\rho_\Gamma^2 \frac{z^2}{H^u(z)} - d_o\rho_\Gamma^3 z^3\}$$

$$\equiv vACD \qquad (5.8.12)$$

$(L_M 501-511)$, and therefore for $\delta R^u(z) = \delta R(\rho_\Gamma z)$

$$\delta R^u(z) = f^u_{acd}(z)(\frac{\delta\Gamma^u(z)}{z^3}) \equiv vDRHS \qquad (5.8.13)$$

(L47). In the remaining lines (L51-62) we calculate first (L52-57)

$$\delta\hat{J}_1(x) = -\hat{F}'(\hat{\Gamma}_p(x))\delta\hat{\Gamma}(x) = [-x \cdot \hat{F}'(\hat{\Gamma}_p(x))] \cdot [\delta\hat{\Gamma}(x)/x] , \qquad (5.8.14)$$

with $\hat{F}(x) \equiv xF\emptyset$, and with $(L_M 515-519)$

$$-x\hat{F}'(\hat{\Gamma}_p(x)) \equiv xDZF\emptyset .$$ (5.8.15)

Remember that only the first $(N - 2) + 1$ coefficients of $\hat{F}(x)$ are known and therefore only the first $(N - 3) + 1$ coefficients of $\hat{F}'(x)$ are known. But since $\delta\hat{\Gamma}(x) = 0(x^k)$, $k \geq 3$, we know the coefficients of $\delta\hat{J}_1(x)$ up to degree N again. (Therefore we may call CDREDUCE (L66-68), CxF$\emptyset$ (L72) and FSUMBOREL (L81) with degree N again.) In L59-60 we calculate

$$\sup_{z\in\Delta} \left| \frac{\delta\Gamma(z)}{(z/\rho^+)^3} \right| \leq \rho_\Gamma \left| \delta\Gamma^u(\frac{\rho^+}{\rho_\Gamma}.) \right|_B \equiv sT1 ,$$ (5.8.16)

and furthermore we bound (L61)

$$\sup_{z\in\Delta} \left| \frac{\Delta\delta J_{1,G}(z)}{(z/\rho^+)^4} \right| \equiv sGDJ1 ,$$ (5.8.17)

and (L62)

$$\sup_{z\in\Delta} \left| \frac{\Delta\delta J_{1,H}(z)}{(z/\rho^+)^{N-3}} \right| \equiv sHDJ1 .$$ (5.8.18)

(See Table 7.8.1 for the definition of $\Delta\delta J_{1,G}$ and $\Delta\delta J_{1,H}$.)
The bound on (5.8.17) and (5.8.18) follows from (5.8.16) once that

$$\sup_{z\in\Delta} \left| \frac{1}{(z/\rho^+)} [(\Delta F'(\Gamma) + (F'(\Gamma) - F'(\Gamma_p)))(z)]_G \right|$$

$$\leq sGDZDF\emptyset$$ (5.8.19)

and

$$\sup_{z \in \delta} \left| \frac{1}{(z/\rho^+)^{N-6}} [(\Delta F'(\Gamma) + (F'(\Gamma) - F'(\Gamma_P)))(z)]_H \right|$$

$$\leq \text{SHDZDF}\emptyset \qquad (5.8.19a)$$

are known. ($\leq$ means that the expression which we calculate on the computer is an upper bound for the left hand side.) We calculate (5.8.19) and (5.8.19a) in a preparatory calculation $(L_M 245-437)$ which we explain now in detail. We first bound $(L_M 248-271)$

$$\sup_{z \in \delta} \left| \frac{1}{(z/\rho^+)^3} [(F'(\Gamma) - F'(\Gamma_P))(z)]_G \right| \leq \text{sGJDZF} \qquad (5.8.20)$$

and

$$\sup_{z \in \delta} \left| \frac{1}{(z/\rho^+)^{N+1}} [(F'(\Gamma) - F'(\Gamma_P))(z)]_H \right| \leq \text{sHJDZF} . \qquad (5.8.21)$$

We proceed similarly as in Section 5.7.4. We have $F(z) = f_\mathcal{L}(z^2)$, and $f_{B,\mathcal{L}}(t)$ the Borel transform of $f_\mathcal{L}$, and therefore

$$\left| F'(\Gamma(z)) - F'(\Gamma_P(z)) \right|$$

$$= \left| \int_0^\infty e^{-t/(\Gamma_P(z))^2} \{ e^{-\eta(z)t} \frac{2t}{(\Gamma(z))^3} - \frac{2t}{(\Gamma_P(z))^3} \} f_{B,\mathcal{L}}(t) dt \right|$$

$$\leq \left| \int_0^\infty e^{-t/(\Gamma_P(z))^2} \frac{2t}{(\Gamma_P(z))^3} ((1 + \frac{\Delta\Gamma(z)}{\Gamma_P(z)})^{-3} - 1) f_{B,\mathcal{L}}(t) dt \right|$$

$$+ \left| \int_0^\infty e^{-t/(\Gamma_P(z))^2} \frac{2t}{(\Gamma(z))^3} (\eta(z)t) (\frac{e^{-\eta(z)t} - 1}{\eta(z) \cdot t}) f_{B,\mathcal{L}}(t) dt \right| ,$$

$$(5.8.22)$$

where $\eta(z)$ is given by (5.7.45). We proceed now exactly as for

(5.7.52). Note that not only the function $\eta(z)$ but also

$$(1+\frac{\Delta\Gamma(z)}{\Gamma_P(z)})^{-3} - 1 = \frac{-3-3\frac{\Delta\Gamma(z)}{\Gamma(z)} - (\frac{\Delta\Gamma(z)}{\Gamma(z)})^2}{(1 + \Delta\Gamma(z)/\Gamma(z))^3} \cdot \frac{\Delta\Gamma(z)}{\Gamma(z)} \qquad (5.8.23)$$

has a factor $\Delta\Gamma(z)$ and we can therefore split the function

$F'(\Gamma(z)) - F'(\Gamma_P(z))$ into a general part which is $0(z^3)$ and a

higher order part which is $0(z^{N+1})$. This has been anticipated

in (5.8.20), (5.8.21). Using this decomposition we get, using

again, where necessary, for $a > 0$,

$$\int_0^\infty t^n e^{-at} dt = \frac{n!}{a^{n+1}}, \qquad (5.8.24)$$

$$\text{sGJDZF} \equiv \{\frac{2}{\eta^4} \frac{3+3X/\eta+ (X/\eta)^2}{(1-X/\eta)^3} C_{\mathcal{L}}(\frac{1}{r} - \frac{(\rho^+)^2}{R_{\mathcal{L}}})^{-2}$$

$$+ \frac{4}{\eta^6} \frac{2+X/\eta}{(1-X/\eta)^5} \cdot C_{\mathcal{L}} \cdot (\frac{1}{r} - YZ(\rho^+)^2 - \frac{(\rho^+)^2}{R_{\mathcal{L}}})^{-3}\}$$

$$\cdot \sup_{z\in\delta} |\frac{\Delta\Gamma_G(z)}{(z/\rho^+)^3}| \qquad (5.8.25)$$

and

$$\text{sHJDZF} \equiv \{\text{same expression as in (5.8.25)}\} \cdot$$

$$\cdot \sup_{z\in\delta} |\frac{\Delta\Gamma_H(z)}{(z/\rho^+)^{N+1}}| \cdot \qquad (5.8.26)$$

These bounds are performed in $L_M 269$ and $L_M 271$ respectively.

This completes our discussion of $L_M 250-271$. In order to get the bounds (5.8.19) and (5.8.19a) we need in addition the bounds

$$\sup_{z \in \delta} \left| \frac{1}{(z/\rho^+)} \Delta F_G'(z) \right| \ , \tag{5.8.27}$$

and

$$\sup_{z \in \delta} \left| \frac{1}{(z/\rho^+)^{N-6}} \Delta F_H'(z) \right| \ . \tag{5.8.28}$$

This turns out to be the most painful estimation of this section and will take up most of it (to Eq. (5.8.72)). It is carried out in $L_M 273-437$. The strategy is the following : By definition (Section 5.7.2), ΔF is the solution of the equation

$$\Delta F(\Gamma) - \Delta F = \Delta J + (F(\Gamma) - F(\Gamma_p)) \ , \tag{5.8.29}$$

where ΔJ is produced by the subroutine FREDUCE, and $(F(\Gamma) - F(\Gamma_p))$ by the subroutine FBNDIM in FADDGENERAL. (See Section 5.7.5.) Taking the derivative of (5.8.29) leads to

$$\Delta F'(\Gamma)\Gamma' - \Delta F' = \Delta J'_{tot} \ , \tag{5.8.30}$$

with

$$\Delta J'_{tot} = \Delta J' + (F'(\Gamma)\Gamma' - F'(\Gamma_p)\Gamma_p')$$

$$= \Delta J' + F'(\Gamma_p)\Delta\Gamma' + (F'(\Gamma) - F'(\Gamma_p))\Gamma' \ . \tag{5.8.31}$$

It is clear that we first have to define the right hand side (5.8.31) of (5.8.30) before we can start to solve the equation, but since our choice of the decomposition of (5.8.31) into general and higher order terms is dictated by the necessity to overcome some additional problems in (5.8.30) we first discuss how

the solution of this equation looks like on a formal level. Dif-

ferentiating the usual representation of the solution ΔF of

(5.8.29) with respect to z leads to

$$\Delta F'(z) = \sum_{n=0}^{\infty} \Delta J'_{tot}(\Gamma^n(z)) \prod_{j=0}^{n-1} \Gamma'(\Gamma^j(z)) . \qquad (5.8.32)$$

Suppose momentarily that $\Delta J'_{tot}(z) = O(z^m)$, and that

$$Q := \sup_{z \in \mathcal{s}} \left| \frac{\Delta J'_{tot}(z)}{(z/\rho^+)^m} \right| < \infty . \qquad (5.8.33)$$

Then we have the estimate

$$|\Delta F'(z)| \le Q \cdot \left(\frac{1}{\rho^+}\right)^m$$

$$\cdot \left\{ |z|^m + \sum_{n=1}^{\infty} |\Gamma^n(z)|^m \cdot \sum_{j=0}^{n-1} |\Gamma'(\Gamma^j(z))| \right\} . \qquad (5.8.34)$$

If we define on the other hand

$$A = \sup_{z \in \mathcal{s}} \left| \frac{\log(\Gamma'(z))}{z^2} \right| . \le \left(\frac{1}{\rho^+}\right)^2 \left| \log\Gamma^{u'}\left(\frac{\rho^+}{\rho_\Gamma}.\right) \right|_B$$

$$\equiv sLOGDZGAMMA \qquad (5.8.35)$$

$(L_M 404\text{-}411)$, then we have

$$|\log(\Gamma'(z))| \le A|z|^2 . \qquad (5.8.36)$$

Using in addition the inequality $\operatorname{Re}((\frac{1}{\Gamma(z)})^2 - \frac{1}{z^2}) \geq L_5 > 0$ (see (5.7.6)), we get

$$| \prod_{j=0}^{n-1} \Gamma'(\Gamma^j(z))| \leq \exp(A \sum_{j=0}^{n-1} |(\Gamma^j(z))^2|)$$

$$\leq \exp(A|z|^2 + A \cdot \int_0^{n-1} \frac{dx}{L_5 x + L_6/|z|^2})$$

$$= \exp(A|z|^2) \cdot (1 + \frac{L_5}{L_6}|z|^2(n-1))^{A/L_5} ,$$

where $L_5 \equiv sL5$, $L_6 \equiv sL6$. Inserting this estimate into (5.8.34) and using (5.8.33) leads to

$$|\Delta F'(z)| \leq Q(\frac{1}{\rho_+})^m \cdot \{|z|^m$$

$$+ \exp(A|z|^2) \cdot [|\Gamma(z)|^m + \sum_{n=2}^{\infty} \frac{(1+\frac{L_5}{L_6}|z|^2(n-1))^{A/L_5}}{(1+\frac{L_5}{L_6}|z|^2 n)^{m/2}} \cdot (\frac{L_6}{|z|^2})^{-m/2}]\}$$

$$\leq Q \cdot (\frac{1}{\rho_+})^m \cdot \{|z|^m + \exp(A|z|^2)[|z|^m$$

$$+ \int_0^{\infty} (1 + \frac{L_5}{L_6}|z|^2 x)^{A/L_5 - m/2} (\frac{|z|^2}{L_6})^{m/2} dx]\}$$

$$\leq Q \cdot (\frac{1}{\rho_+})^m \cdot \{|z|^m + \exp(A|z|^2)[|z|^m$$

$$+ (\frac{|z|^2}{L_6})^{m/2-1} \frac{1}{L_5} \frac{1}{m/2 - A/L_5 - 1}]\} \qquad (5.8.37)$$

From this estimate it follows that the sum (5.8.32) defining $\Delta F'$ converges if $m/2 - 1 - A/L_5 > 0$. Unfortunately $A/L_5 \gtrsim 3/2$ for the problem at hand and m has to be at least 6 for the sum

(5.8.32) to be convergent. This poses a problem since $\Delta J_G = 0(z^4)$ (see (5.7.55)) and therefore the derivative of ΔJ_G would be $0(z^3)$. We are therefore forced to treat the orders $0(z^3) - 0(z^5)$ separately, i.e. we have to split the lower orders into explicit orders $0(z^3) - 0(z^5)$ and a general function of order $0(z^6)$.

This being said, we have now to define a decomposition of the function $\Delta J'_{tot} = \Delta J' + F'(\Gamma_p)\Delta\Gamma' + (F'(\Gamma) - F'(\Gamma_p))\Gamma'$. We do this as follows : We discuss in a first step the higher order term, called $J'_{tot,H}$ and in a second step the explicit terms and the general term. We define $\Delta J'_{tot,H}$ by the equation

$$\Delta J'_{tot,H} = (\Delta J_H)' + F'(\Gamma_p)(\Delta\Gamma_H)'$$

$$+ [F'(\Gamma) - F'(\Gamma_p)]_H \hat{}\Gamma' , \tag{5.8.38}$$

with $(\Delta J_H)'$ as defined in (5.8.3). The definition of the expression $[\ ]_H \hat{}$ will be given now. In principle we have already a similar expression $[\ ]_H$ and the bound sHJDZF (5.8.21) at our disposal, but the modification of the corresponding general term $[\ ]_G$ (5.8.20), which is necessary to make it $0(z^6)$ makes also a new definition of the higher order term necessary. We call this higher order term $[\ ]_H \hat{}$ and we get it as follows. Let again $f_{B,\mathcal{L}}(t)$ be the Borel transform of $f_{\mathcal{L}}(z)$, where $F(z) = f_{\mathcal{L}}(z^2)$. We define then $\tilde{f}_{B,\mathcal{L}}$ by the equation

$$\tilde{f}_{B,\mathcal{L}}(t) = f_{B,\mathcal{L}}(t) - B_0 - B_1 t^{1/2} - B_2 t , \tag{5.8.39}$$

where B_0, B_1 and B_2 are the first three coefficients of the expansion of $f_{B,\mathcal{L}}$ at $t = 0$, and therefore $\tilde{f}_{B,\mathcal{L}}(t) = 0(t^{3/2})$.

For $t > 0$ we have the inequality $t < a \cdot \exp(t/(ae))$, $a > 0$ and $t^{1/2} < b \exp(t/(2b^2 e))$, $b > 0$. We choose $a = R/e$ and $b = (R/(2e))^{1/2}$ and get for $t \in S_{2,\beta}$

$$|\tilde{f}_{B,\mathcal{L}}(t)| \leq \tilde{C}_{\mathcal{L}} \exp(|t|/R_{\mathcal{L}}) ,\qquad (5.8.40)$$

where $(L_M 285\text{-}287)$

$$\tilde{C}_{\mathcal{L}} = C + |B_0| + |B_1|(\frac{R}{2e})^{1/2} + |B_2|(\frac{R}{e}) \equiv sCT . \qquad (5.8.41)$$

Furthermore we have $(L_M 282\text{-}284)$ $B_0 \equiv sB0$, $B_1 \equiv sB1$, $B_2 \equiv sB2$. In addition to the bound (5.8.40) for $\tilde{f}_{B,\mathcal{L}}(t)$ we shall need a bound which shows explicitly that $\tilde{f}_{B,\mathcal{L}}(t) = 0(t^{3/2})$, i.e. we have to find a constant $\tilde{\tilde{C}}_{\mathcal{L}}$, such that $t \in S_{2,\beta}$

$$|\tilde{f}_{B,\mathcal{L}}(t)| \leq \tilde{\tilde{C}} \cdot |t|^{3/2} \cdot \exp(|t|/R_{\mathcal{L}}) . \qquad (5.8.42)$$

Let $\partial S_{2,\beta}$ be the boundary of the domain $S_{2,\beta}$. For $t \in \partial S_{2,\beta}$ we have that $|t| \geq r_0 = \beta \log 2$, i.e. the point $-\beta \log 2 \in \partial S_{2,\beta}$ is the point nearest to the origin $t = 0$. It follows that $S_{2,\beta}$ contains the disk $D_2(0,r_0)$. Define now the domains $S_{2,\beta}(r)$ by $S_{2,\beta}(r) = \{t \in S_{2,\beta} \mid |t| < r\}$. By construction we have that $\tilde{f}_{B,\mathcal{L}}(t) = 0(t^{3/2})$ near $t = 0$ and therefore $f_{B,\mathcal{L}}/t^{3/2}$ is analytic in $S_{2,\beta}$. Let now $|t| > r_0$, then

$$\left|\frac{\tilde{f}_{B,\mathcal{L}}(t)}{t^{3/2}}\right| \le \sup_{\tau \in S_{2,\beta}(|t|)} \left|\frac{\tilde{f}_{B,\mathcal{L}}(\tau)}{\tau^{3/2}}\right|$$

$$\le (\frac{1}{r_o})^{3/2} \tilde{C}_{\mathcal{L}} \exp(|t|/R_{\mathcal{L}}) \ . \tag{5.8.43}$$

For $|t| < r_o$ we have

$$\left|\frac{\tilde{f}_{B,\mathcal{L}}(t)}{t^{3/2}}\right| \le \sup_{t \in \partial D(0,r_o)} \left|\frac{f_{B,\mathcal{L}}(t)}{t^{3/2}}\right|$$

$$\le \tilde{C}_{\mathcal{L}}(\frac{1}{r_o})^{3/2} \cdot \exp(\frac{r_o}{R_{\mathcal{L}}}) \le \tilde{C}_{\mathcal{L}}(\frac{1}{r_o})^{3/2} \cdot \exp(\frac{r_o}{R_{\mathcal{L}}}) \exp(|t|/R_{\mathcal{L}})$$

$$\tag{5.8.44}$$

and therefore if we choose $(L_M 288-290)$

$$\tilde{\tilde{C}}_{\mathcal{L}} = (\frac{1}{r_o})^{3/2} \exp(\frac{r_o}{R_{\mathcal{L}}}) \tilde{C}_{\mathcal{L}} \equiv \text{sCTT} \ , \tag{5.8.45}$$

then the inequality (5.8.42) is satisfied for $t \in S_{2,\beta}$. Using now the definition of $\tilde{f}_{B,\mathcal{L}}$ we get

$$F(z) = \int_0^\infty e^{-t/z^2}(\tilde{f}_{B,\mathcal{L}}(t) + B_o + B_1 t^{1/2} + B_2 t) dt$$

$$= \frac{1}{2}F_o z^2 + \frac{1}{3}F_1 z^3 + \frac{1}{4}F_2 z^4 + \hat{F}(z) \ , \tag{5.8.46}$$

where $F_o = 2B_o$, $F_1 = \frac{3}{2}\pi^{1/2}B_1$, $F_2 = 4B_2$ and $\hat{F}(z)$ is given by

$$\hat{F}(z) = \int_0^\infty e^{-t/z^2} \tilde{f}_{B,\mathcal{L}}(t) dt \ , \tag{5.8.47}$$

and $\hat{F}(z) = O(z^5)$ as is easy to see, using the bound (5.8.42)

for $\tilde{f}_{B,\mathcal{L}}$. Using this decomposition for $F(z)$ we can define

$$[F'(\Gamma) - F'(\Gamma_p)]\hat{}_H = \{F_o + F_1(2\Gamma_p + \Delta\Gamma)$$

$$+ F_2(3\Gamma_p^2 + 3\Gamma_p\Delta\Gamma + (\Delta\Gamma)^2)\}\Delta\Gamma_H$$

$$+ [\hat{F}'(\Gamma) - \hat{F}'(\Gamma_p)]_H \ . \tag{5.8.48}$$

For the definition of the expression $[\hat{} \ \hat{}]_H$ we proceed now exactly as in the calculation of this expression for F instead of $\hat{F}$ (see (5.8.20)-(5.8.25)) and we use also the same decomposition into general parts and higher order parts.

The function $\Delta J'_{tot,H}$, (5.8.38), is now well-defined and we have the bound $(L_M 367\text{-}373)$

$$\sup_{z\in\delta}\left|\frac{\Delta J'_{tot,H}(z)}{(z/\rho^+)^{N-3}}\right| \leq \sup_{z\in\delta}\left|\frac{\Delta J'_H(z)}{(z/\rho^+)^{N-3}}\right|$$

$$+ \sup_{z\in\delta}|F'(\Gamma_p(z))|\cdot\sup_{z\in\delta}\left|\frac{(\Delta\Gamma_H)'(z)}{(z/\rho^+)^N}\right|$$

$$+ \sup_{z\in\delta}\left|\frac{1}{(z/\rho^+)^{N+1}}[F'(\Gamma) - F'(\Gamma_p)]\hat{}_H(z)\right|\cdot\sup_{z\in\delta}|\Gamma'(z)|$$

$$\equiv \text{sHDZJTOT} \ . \tag{5.8.49}$$

A bound for $(\Delta J_H)' \equiv \text{sHDZJ}$, has already been given in (5.8.4).

We give now first the bound $(L_M 301)$

$$\sup_{z \in \mathfrak{s}} \left| \frac{(\Delta \Gamma_H)'(z)}{(z/\rho^+)^N} \right| \leq \left| (\Delta \Gamma_H^u)'(\frac{\rho^+}{\rho_\Gamma}\cdot) \right|_1 \equiv \text{SHDZGAMMA} \tag{5.8.50}$$

and the bound $(L_M 310\text{-}312)$

$$\sup_{z \in \mathfrak{s}} |\Gamma'(z)| \leq \left| \Gamma^{u'}(\frac{\rho^+}{\rho_\Gamma}\cdot) \right|_B \equiv \text{SDZGAMMA} . \tag{5.8.51}$$

We proceed to bound $\sup_{z \in \mathfrak{s}} |F'(\Gamma_p(z))|$. We have

$$\sup_{z \in \mathfrak{s}} \left| \frac{F'(\Gamma_p(z))}{z/\rho^+} \right| = \left| \frac{1}{z/\rho^+} \cdot \int_0^\infty e^{-t/(\Gamma_p(z))^2} \frac{2t}{(\Gamma_p(z))^3} f_{B,\mathcal{L}}(t) dt \right|$$

$$\leq 2 \cdot C_{\mathcal{L}} \cdot \frac{\rho^+}{\eta^3} (\frac{1}{r} - \frac{(\rho^+)^2}{R_{\mathcal{L}}})^{-2} \equiv \text{sDZF0GAMMA} \tag{5.8.52}$$

$(L_M 294\text{-}295)$, and it remains to bound $[\ ]_H^\wedge$. We decompose F as described above, take its derivative, $F'(z) = F_0 z + F_1 z^2 + F_2 z^3 + \hat{F}'(z)$, $\hat{F}'(z) = O(z^4)$, i.e. we use (5.8.48) for $[\ ]_H^\wedge$ and get

$$\sup_{z \in \mathfrak{s}} \left| \frac{1}{(z/\rho^+)^{N+1}} [F'(\Gamma) - F'(\Gamma_p)]_H^\wedge(z) \right|$$

$$\leq \{ |F_0| + |F_1| \cdot \{ \sup_{z \in \mathfrak{s}} |\Gamma(z)| + \sup_{z \in \mathfrak{s}} |\Gamma_p(z)| \}$$

$$+ |F_2| \cdot \{ 3 \cdot \sup_{z \in \mathfrak{s}} |(\Gamma_p(z))^2| + 3 \cdot \sup_{z \in \mathfrak{s}} |\Gamma_p(z)| \cdot \sup_{z \in \mathfrak{s}} |\Delta\Gamma(z)|$$

$$+ (\sup|\Delta\Gamma|)^2\}\} \cdot \sup_{z\in\mathcal{S}} \left| \frac{\Delta\Gamma_H}{(z/\rho^+)^{N+1}} \right|$$

$$+ \sup_{z\in\mathcal{S}} \left| \frac{1}{(z/\rho^+)^{N+1}} [\hat{F}'(\Gamma) - \hat{F}'(\Gamma_P)]_H(z) \right| . \tag{5.8.53}$$

We have $(L_M 275\text{-}279)$: $F_O \equiv sFP00$, $F_1 \equiv sFP01$, $F_2 \equiv sFP02$ and furthermore $(L_M 305\text{-}306)$

$$\left.\begin{array}{c} \sup_{z\in\mathcal{S}}|\Gamma(z)| \\[2em] \sup_{z\in\mathcal{S}}|\Gamma_P(z)| \end{array}\right\} \leq \rho_\Gamma \left| \Gamma^u(\frac{\rho^+}{\rho_\Gamma}.) \right|_B \equiv sGAMMA , \tag{5.8.54}$$

and $(L_M 316\text{-}318)$

$$\sup_{z\in\mathcal{S}} |(\Gamma_P(z))^2| \leq \rho_\Gamma^2 \left| (\Gamma^u(\frac{\rho^+}{\rho_\Gamma}.))^2 \right|_B \equiv s2GAMMA . \tag{5.8.55}$$

all the other bounds have already been established, with the exception of $[\overset{\wedge}{\quad}]_H$, which by construction of the function $\hat{F}_1$ and using the bound (5.8.40) for $\tilde{f}_{B,\mathcal{L}}$, is given by

$$\sup_{z\in\mathcal{S}} \left| \frac{1}{(z/\rho^+)^{N+1}} [\hat{F}'(\Gamma) - \hat{F}'(\Gamma_P)]_H(z) \right|$$

$$\leq \frac{\tilde{C}_\mathcal{L}}{C_\mathcal{L}} \cdot \sup_{z\in\mathcal{S}} \left| \frac{1}{(z/\rho^+)^{N+1}} [F'(\Gamma) - F'(\Gamma_P)]_H(z) \right| \tag{5.8.56}$$

where the supremum in the right hand side has already been bounded in (5.8.21). We have given bounds for all the ingredients of (5.8.49), and this bound is implemented in $L_M 367\text{-}373$. This completes our discussion of the higher order term of $\Delta J'_{tot}$, (5.8.38), and we complete now the discussion by giving bounds on the general term. But as indicated above we have to take care and split terms up to orders $O(z^5)$, i.e. we want to write

$$\Delta J'_{tot}(z) = J_1 z^3 + J_2 z^4 + J_3 z^5 + \Delta J'_{tot,G}(z) + \Delta J'_{tot,H}(z) ,$$

$$(5.8.57)$$

with $\Delta J'_{tot,H}$ as defined above and with $\Delta J'_{tot,G}(z) = O(z^6)$. We already have decomposed $\Delta J'$ (see (5.8.3)-(5.8.8)), namely we have for $z \in D(0,\rho^+)$

$$\Delta J'(z) = J^o_1 z^3 + J^o_2 z^4 + J^o_3 z^5 + \Delta J'_{tot,G} + \Delta J'_{tot,H} ,$$

and we need a similar decomposition for the function

$$F'(\Gamma_p(z)) (\Delta \Gamma_G)'(z) , \qquad (5.8.58)$$

and for the function

$$\{F_o + F_1 \cdot (2\Gamma_p + \Delta\Gamma) + F_2 (3(\Gamma_p)^2 + 3\Gamma_p \cdot \Delta\Gamma + (\Delta\Gamma)^2)\}$$

$$\cdot \Gamma' \cdot \Delta\Gamma_G + [\hat{F}'(\Gamma) - \hat{F}'(\Gamma_p)]_G \cdot \Gamma' . \qquad (5.8.59)$$

By construction, $\Delta\Gamma_G$ is of the form

$$\Delta\Gamma_G(z) = az^3 + bz^5 + O(z^6) ,$$

where $(|a| (\rho_\Gamma)^2) (\rho_\Gamma \cdot (\frac{\rho^+}{\rho_\Gamma})^3) \leq$ sGGAMMA, and $(|b| (\rho_\Gamma)^4)$

$(\rho_\Gamma \cdot (\frac{\rho^+}{\rho_\Gamma})^3) \leq$ sGGAMMA. Furthermore, there is a constant κ_P for

which $\Gamma_P(z) = z - (2/\kappa_P) z^3 + O(z^5)$. Inserting now the decomposi-

tion (5.8.46) into (5.8.58) leads to

$$\{F_0 \cdot \Gamma_P(z) + F_1 (\Gamma_P(z))^2 + F_2 (\Gamma_P(z))^3 + \hat{F}'(\Gamma_P(z))\}(\Delta\Gamma_G)'(z)$$

$$= \{F_0 \cdot (z - (2/\kappa_P) z^3) + F_1 z^2 + F_2 z^3\}(3az^2 + 5bz^4) + O(z^6)$$

$$= 3aF_0 z^3 + 3aF_1 z^4 + \{(-\frac{6a}{\kappa_P} + 5b)F_0 + 3aF_2\}z^5 + O(z^6) ,$$

and we have used that $\hat{F}'(z) = O(z^4)$, which again follows using

the bound (5.8.42) for $\tilde{f}_{B,\mathcal{L}}$. Next we consider (5.8.59). Note

that $[\hat{F}'(\Gamma) - \hat{F}'(\Gamma_P)] = O(z^6)$ and therefore we have to extract

in (5.8.59) coefficients only from the other terms. We have

$$\{F_0 + F_1 \cdot 2z + F_2 \cdot 3z^2\}(1 - \frac{6}{\kappa_P}z^2)(az^3 + bz^5) + O(z^6)$$

$$= F_0 az^3 + F_1 \cdot 2az^4 + \{F_0(-\frac{6}{\kappa_P}a + b) + F_2 \cdot 3a\}z^5 + O(z^6) .$$

Collecting coefficients leads us to define J_1, J_2 and J_3 in

(5.8.57) as $J_1 = J_1^o + \hat{J}_1 \equiv$ sJ1, $(L_M 345)$, $J_2 = J_2^o + \hat{J}_2 \equiv$ sJ2,

$(L_M 347)$, $J_3 = J_3^o + \hat{J}_3 \equiv$ sJ3, $(L_M 351)$ where

$$\hat{J}_1 = 4aF_0 \equiv \text{sJ1H} \tag{5.8.60}$$

$$\hat{J}_2 = 5aF_1 \equiv \text{sJ2H} \tag{5.8.61}$$

$$\hat{J}_3 = (-\frac{12a}{\kappa_P} + 6b)F_0 + 6aF_2 \equiv sJ3H \qquad (5.8.62)$$

($L_M 344$, $L_M 346$ and $L_M 348-350$, respectively). We çan now define the general term $\Delta J'_{tot,G}$ by the equation

$$\Delta J'_{tot,G}(z) = \Delta J'_G(z) + (\sum_{i=0}^{2} F_i \cdot (\Gamma_P(z))^{i+1})(\Delta\Gamma)'(z)$$

$$+ \hat{F}'(\Gamma_P(z))(\Delta\Gamma_G)'(z) + \{F_0 + F_1 \cdot (2\Gamma_P(z) + \Delta\Gamma(z))$$

$$+ F_2 \cdot 3((\Gamma_P(z))^2 + 3\Gamma_P(z)\Delta\Gamma(z) + (\Delta\Gamma(z))^2)\}$$

$$\cdot \Gamma'(z)\Delta\Gamma_G(z) + [\hat{F}'(\Gamma) - \hat{F}'(\Gamma_P)]_G(z)\cdot\Gamma'(z)$$

$$- \sum_{i=1}^{3} \hat{J}_i \cdot z^{i+2} . \qquad (5.8.63)$$

By construction, $\Delta J'_{tot}(z) = 0(z^6)$. Using for that part of (5.8.63) which is an analytic function not only on δ, but even on $D(0,\rho^+)$ (all terms not involving $\hat{F}'$) the fact that it takes its maximal value on $\partial D(0,\rho^+)$ we can bound $\Delta J'_{tot,G}$ by

$$\sup_{z\in\delta}|\frac{\Delta J'_{tot,G}}{(z/\rho^+)^6}| \leq \sup_{z\in\delta}|\frac{\Delta J'_G(z)}{(z/\rho^+)^6}|$$

$$+ \{\sum_{i=0}^{2}|F_i| \cdot \sup_{z\in\partial D(0,\rho^+)}|(\Gamma_P(z))^{i+1}|\} \cdot \sup_{z\in\partial D(0,\rho^+)}|(\Delta\Gamma_G)'(z)|$$

$$+ \{|F_0| + |F_1| \cdot (2 \cdot \sup_{z\in\partial D(0,\rho^+)}|\Gamma_P(z)| + \sup_{z\in\partial D(0,\rho^+)}|\Delta\Gamma(z)|)$$

$$+ |F_2| \cdot (3 \cdot \sup_{z \in \partial D(0,\rho^+)} |(\Gamma_p(z))^2|$$

$$+ 3 \cdot \sup_{z \in \partial D(0,\rho^+)} |\Gamma_p(z)| \cdot \sup_{z \in \partial D(0,\rho^+)} |\Delta\Gamma(z)|$$

$$+ \sup_{z \in \partial D(0,\rho^+)} |(\Delta\Gamma(z))^2|)\}$$

$$\cdot \sup_{z \in \partial D(0,\rho^+)} |\Gamma'(z)| \cdot \sup_{z \in \partial D(0,\rho^+)} |\Delta\Gamma_G(z)| + \sum_{i=1}^{3} |\hat{J}_i|(\rho^+)^{i+2}$$

$$+ \sup_{z \in \delta} |\frac{1}{(z/\rho^+)^6}[\hat{F}'(\Gamma) - \hat{F}'(\Gamma_p)]_G(z)| \cdot \sup_{z \in \delta} |\Gamma'(z)|$$

$$+ \sup_{z \in \delta} |\frac{1}{(z/\rho^+)^4} \hat{F}'(\Gamma_p(z))| \cdot \sup_{z \in \delta} |\frac{1}{(z/\rho^+)^2}(\Delta\Gamma_G)'(z)|$$

$$\equiv \text{sGDZJTOT} .$$
(5.8.64)

$(L_M 353\text{-}364)$. We have already established bounds for all the ingredients of (5.8.64) with the exception of $(L_M 300)$

$$\sup_{z \in \delta} |\frac{(\Delta\Gamma_G)'(z)}{(z/\rho^+)^2}| \leq |(\Delta\Gamma_G^u)'(\frac{\rho^+}{\rho_\Gamma} \cdot)|_1 \equiv \text{sGDZGAMMA} ,$$

and those for terms involving $\hat{F}'$. Using (5.8.52) with $\tilde{\tilde{f}}_{B,\mathcal{L}}$ instead of $f_{B,\mathcal{L}}$, the bound (5.8.42) and furthermore (5.8.24) leads to $(L_M 327\text{-}336, L_M 338\text{-}339)$

$$\sup_{z \in \delta} |\frac{\hat{F}'(\Gamma_p(z))}{(z/\rho^+)^4}| \leq 2\tilde{c}_\mathcal{L}(\frac{5}{2})!(\rho^+)^4 \frac{1}{\eta^3}(\frac{1}{r} - \frac{(\rho^+)^2}{R_\mathcal{L}})^{-7/2}$$

$$\equiv \text{sTDZF\O GAMMA}$$
(5.8.65)

and furthermore using (5.8.22) with the same replacements leads

to $(L_M 337)$

$$\sup_{z \in \delta} \left| \frac{1}{(z/\rho^+)^6} [\hat{F}'(\Gamma) - \hat{F}'(\Gamma_p)]_G(z) \right|$$

$$\leq \left\{ \frac{2}{\eta^4} \frac{3 + 3X/\eta + (X/\eta)^2}{(1-X/\eta)^3} \left(\frac{5}{2}\right) : \tilde{C}_\mathcal{L} \left(\frac{1}{r} - \frac{(\rho^+)^2}{R_\mathcal{L}}\right)^{-7/2} \right.$$

$$+ \frac{2}{\eta^6} \frac{2 + X/\eta}{(1-X/\eta)^5} \cdot \left(\frac{7}{2}\right) : \tilde{C}_\mathcal{L} \left(\frac{1}{r} - YZ(\rho^+)^2 - \frac{(\rho^+)^2}{R_\mathcal{L}}\right)^{-9/2}$$

$$\cdot (\rho^+)^3 \cdot \sup_{z \in \delta} \left| \frac{\Delta\Gamma_G(z)}{(z/\rho^+)^3} \right| \equiv \text{sTGJDZF} \tag{5.8.66}$$

This completes our discussion of the definition of the right hand

side in (5.8.31), i.e. $\Delta J'_{tot}$ is now defined and bounded on δ.

We next want to solve for $\Delta F'$ in (5.8.30) and we solve now first

for the explicitly known terms in order to achieve for $m \geq 6$ in

(5.8.37) in order to make that series convergent. We make the an-

satz

$$\Delta F'(z) = \Delta F_1 z + \Delta F_2 z^2 + \Delta F_3 z^3 + \Delta\hat{F}'(z) , \tag{5.8.67}$$

and get by substitution into (5.8.30) :

$$\Delta\hat{F}'(\Gamma(z))\Gamma'(z) - \Delta\hat{F}'(z)$$

$$= \Delta J'_{tot,G}(z) + \Delta J'_{tot,G,additional}(z) + \Delta J'_{tot,H}(z) = 0(z^6) ,$$

where $\Delta J'_{tot,G,additional}$ is given by the equation

$$\Delta J'_{tot,G,additional}(z) = \sum_{i=1}^{3} J_i z^{i+2}$$

$$- \sum_{i=1}^{3} \Delta F_i \{(\Gamma(z))^i \Gamma'(z) - z^i\} , \qquad (5.8.68)$$

with $\Delta F_1 = J_1/4\gamma_3 \equiv sDZF01$, $\Delta F_2 = J_2/5\gamma_3 \equiv sDZF02$ and

$\Delta F_3 = 1/6\gamma_3 (J_3 - 6\Delta F_1 \cdot \gamma_5 - \frac{3}{4} J_1 \gamma_3) \equiv sDZF03$ in $L_M 375$, $L_M 376$ and

$L_M 377-379$ respectively, and where $\gamma_3 \equiv \frac{1}{6}\Gamma'''(0) \equiv sGA3$ and

$\gamma_5 \equiv \frac{1}{5!}\Gamma^{(v)}(0) \equiv sGA5$ (calculated in L53 and L55-56 of FvKI

directly from G_o in order to gain precision).

To bound $\Delta J'_{tot,G,additional}$ we have again to use the fact
that we deal with a function which is analytic on $D(0,\rho^+)$ and
not only on δ. We get

$$\sup_{z \in \delta} |\frac{1}{(z/\rho^+)^6} \Delta J'_{tot,G,additional}(z)| \leq \sum_{i=1}^{3} |J_i|(\rho^+)^{i+2}$$

$$+ \sum_{i=1}^{3} |\Delta F_i| \cdot \sup_{z \in \partial D(0,\rho^+)} |(\Gamma(z))^i \Gamma'(z) - z^i| . \qquad (5.8.69)$$

We decompose now $\Delta \hat{F}' = \Delta \hat{F}'_G + \Delta \hat{F}'_H$ in the evident way, namely
$\Delta \hat{F}'_G$ and $\Delta \hat{F}'_H$ are defined to solve Ecalle's equation (5.8.30)
with only general respectively higher order inhomogeneities. These
inhomogeneities are now $0(z^6)$ and $0(z^{N-3})$, respectively, and
we can use the bound (5.8.37) to establish $(L_M 421-423)$

$$\sup_{z \in \delta} |\frac{\Delta \hat{F}'_G(z)}{(z/\rho^+)^4}| \leq (\frac{1}{\rho^+})^6 \{(\rho^+)^6 + \exp(A(\rho^+)^2) [(\rho^+)^6$$

$$+ \ (\frac{(\rho^+)^2}{L_6}) \ \frac{1}{L_5} \ \frac{1}{2-A/L_5}]\}$$

$$\cdot \ \{\sup_{z \in \Delta} |\frac{\Delta J'_{tot,G}}{(z/\rho^+)^6}| \ + \ \sup_{z \in \Delta} |\frac{\Delta J'_{tot,G,additional}}{(z/\rho^+)^6}|\}$$

$$\equiv \ sGREGDZF0 \qquad\qquad (5.8.70)$$

and, using the bound (5.8.49) for $\Delta J'_{tot,H}$ (L_M425-428)

$$\sup_{z \in \Delta} |\frac{\Delta \hat{F}'_H(z)}{(z/\rho^+)^{N-5}}| \ \le \ (\frac{1}{\rho^+})^{N-3}\{(\rho^+)^{N-3} \ + \ \exp(A(\rho^+)^2)[(\rho^+)^{N-3}$$

$$+ \ (\frac{(\rho^+)^2}{L_6})^{(N-5)/2} \frac{1}{L_5} \ \frac{1}{\frac{N-5}{2}-A/L_5}]\}\cdot\sup_{z \in \Delta} |\frac{\Delta J'_{tot,H}}{(z/\rho^+)^{N-3}}|$$

$$\equiv \ sHREGDZF0 \ . \qquad\qquad (5.8.71)$$

Finally, we define the functions $\Delta F'_G$ and $\Delta F'_H$ of the decomposition of $\Delta F'$ in the obvious way, namely

$$\Delta F'_G \ = \ \Delta F_1 \cdot z \ + \ \Delta F_2 \cdot z^2 \ + \ \Delta F_3 \cdot z^3 \ + \ \Delta \hat{F}'_G(z) \quad \text{and} \quad \Delta F'_H \ = \ \Delta \hat{F}'_H.$$ We have the bound

$$\sup_{z \in \Delta} |\frac{\Delta F'_G(z)}{(z/\rho^+)}| \ \le \ |F_1|\rho^+ \ + \ |\Delta F_2| \ (\rho^+)^2 \ + \ |\Delta F_3| \ (\rho^+)^3$$

$$+ \ \sup_{z \in \Delta} |\frac{\Delta \hat{F}'_G(z)}{(z/\rho^+)^4}| \ , \qquad\qquad (5.8.72)$$

and $\Delta F'_H$ satisfies the bound (5.8.71) by definition.

This completes the painful discussion of the definition of

the right hand side for Ecalle's equation occurring in the definition of the tangent map DK (see Table 5.8.1) : Adding (5.8.20)
and (5.8.27) and multiplying with (5.8.16) gives (5.8.17) and adding (5.8.21) and (5.8.28) and multiplying with (5.8.16) gives
(5.8.18). This is done in L_M 433-435 L_M437 and in L59-62 of
the subroutine FvDKI. So, after this lengthy preparatory step,
we can now continue the discussion of the subroutine FvDKI in
much the same way as for FvKI. In a next step we have to discuss the subroutines CDREDUCE and FDREDUCE, where in analogy
to CREDUCE and FREDUCE in FvKI, the solution of Ecalle's
equation is established as a power series together with bounds on
its Borel transform. To get started, we give first the listings
of the subroutines CDREDUCE and FDREDUCE.

CDREDUCE

```
    1:          SUBROUTINE CDREDUCE(xDJ1,IDEG,sGDJ1,sHDJ1,vDRHS,sDA0,sDC0,sDD0,xDJ
    2:        *                    ,sGDJ,sHDJ
    3:        *                    ,vDGAMMA,sCDF0,sEDF0)
    4:          INCLUDE BIGN.STUFF
    5:          COMMON/ECALLE/sC3
    6:          COMMON/DOMAIN/sRHOG0,sRHOGAMMA,sX0h,sCENTh,sRHOh,sCENTG,sRHOG
    7:        *              ,sRHOhB,sRHOPhB,sX0hB,sPHI0,sxKAPPA,sKAPPA
    8:        *              ,sRHOP,sRHO1
    9:          COMMON/SWITCH/ITEST
   10:          INCLUDE BIGN.STRES1
   11:          DIMENSION xDJ1(1),vDRHS(1),xDJ(1),vDGAMMA(1)
   12:          DIMENSION vxDJ2(NDEG)
   13:          DIMENSION  xDJ2(NDEG)
   14:          DIMENSION vT1(NDEG)
   15:
   16:
   17:C Reduction to regular inhomogeneous part
   18:C =======================================
   19:
   20:          CALL FDREDUCE(vDRHS,vA0,vC0,vD0,sDA0,sDC0,sDD0,vxDJ2,sGDJ2,sHDJ2)
   21:
   22:
   23:C Produce unnormalized power series of vxDJ2 and add xDJ1
   24:C ======================================================
   25:
   26:C Produce power series
   27:C --------------------
   28:          CALL CvTOxe(vxDJ2,xDJ2,eT1)
   29:          ITEST=0
   30:          CALL FsDILAT(xDJ2,sINV(sRHOGAMMA),xDJ2)
   31:          ITEST=1
   32:
   33:C Reduce degree
   34:C -------------
   35:          NSAVE=N
   36:          CALL DEG(IDEG)
   37:          CALL CCLEAN(xDJ2)
   38:
   39:C Add regular part xDJ1
   40:C ---------------------
   41:          CALL FEQUAL(xDJ1,xDJ)
   42:          CALL FADD(xDJ2,xDJ)
   43:
   44:C Initialize back to saved degree
   45:C -------------------------------
   46:          CALL DEG(NSAVE)
   47:
   48:
   49:C Calculate bounds sGDJ and sHDJ
   50:C ==============================
   51:
   52:          sGDJ=sUPPER(sSUM(sGDJ1,sGDJ2))
   53:          sHDJ=sUPPER(sSUM(sHDJ1,sHDJ2))
   54:
   55:
   56:C Calculate bounds of Borel function
   57:C ==================================
   58:
   59:C Compute bounds on vxDJ1H
   60:C ------------------------
   61:C..compute bounds for vDELTA
   62:          sEDEL=sUPPER(sQUOT(sxKAPPA,sPROD(sPROD(sRHOGAMMA,sRHOGAMMA)
   63:        *      ,sFOUR)))
   64:          rT1=rL1NORM(vDGAMMA)
   65:          sCDEL=sUPPER(sPROD(sEDEL,sCONST(rT1)))
```

CDREDUCE

```
 66:C..compute bounds for vDZDELTA
 67:       sEDELP=sLOWER(sSUM(sSUM(sEQ,sEDEL),sEXP(sNEG(sONE))))
 68:       sRHO=sSQRT(sQUOT(sxKAPPA,sPROD(sFOUR,sEDELP)))
 69:       CALL FsDILAT(vDGAMMA,sQUOT(sRHO,sRHOGAMMA),vT1)
 70:       vT1(4)=sZERO
 71:       rT1=rL1NORM(vT1)
 72:       sCDELP=sUPPER(sPROD(sPROD(sEDELP,sEDELP),sCONST(rT1)))
 73:C..compute bounds for vDJ1H
 74:       sEDJ1H=sUPPER(sSUM(sSUM(sEFO,sEQ),sEDEL)
 75:      *         ,sQUOT(sTHREE,sEXP(sONE))))
 76:       sT1=sPROD(sSQRT(sQUOT(sFOUR,sxKAPPA)),vDGAMMA(4))
 77:       sCDJ1H=sUPPER(sPROD(sPROD(sTWO,sSUM(sCFO,sCJ))
 78:      *        ,sSUM(sSUM(sT1,sPROD(sHALF,sSUM(sCDELP,sPROD(s3HALF
 79:      *        ,sCDEL)))),sPROD(sCQ,sQUOT(sCDEL,siCONST(6))))))
 80:
 81:C  Compute bounds on vxDJ2H
 82:C  ------------------------
 83:       sEDJ2H=sEDJ1H
 84:       sRHO=sSQRT(sQUOT(sxKAPPA,sPROD(sFOUR,sEDJ2H)))
 85:       CALL FsDILAT(vxDJ2,sQUOT(sRHO,sRHOGAMMA),vT1)
 86:       rT1=rL1NORM(vT1)
 87:       sCDJ2H=sUPPER(sPROD(sPROD(sEDJ2H,sEDJ2H),sCONST(rT1)))
 88:
 89:C  Compute bounds for vDJH
 90:C  -----------------------
 91:       sEDJH=sEDJ1H
 92:       sCDJH=sUPPER(sSUM(sCDJ1H,sCDJ2H))
 93:
 94:C  Compute bounds on vI2ETA (on new domain)
 95:C  ----------------------------------------
 96:       sEI2ETA=sEDJH
 97:       sRHO=sSQRT(sQUOT(sxKAPPA,sPROD(sFOUR,sEI2ETA)))
 98:       CALL FsDILAT(vI2ETA,sQUOT(sRHO,sRHOGAMMA),vT1)
 99:       vT1(1)=sZERO
100:       vT1(2)=sZERO
101:       vT1(3)=sZERO
102:       vT1(4)=sZERO
103:       rT1=rL1NORM(vT1)
104:       sCI2ETA=sUPPER(sPROD(sPROD(sEI2ETA,sEI2ETA),sCONST(rT1)))
105:
106:C  Compute bounds on vDFO
107:C  ----------------------
108:       sC31=sSUM(sC3,sONE)
109:       sT1=sPROD(sCI2ETA,sC31)
110:C..compute sEDFO
111:       rT1=rMAX2(rMAXABS(sT1),rONE)
112:       sEDFO=sUPPER(sSUM(sEDJH,sCONST(rT1)))
113:C..compute sCDFO
114:       sT2=sPROD(sC3,sCDJH)
115:       sT3=sQUOT(sPROD(sC3,sCDJH),sPROD(sT1,sC31))
116:       sT4=sSUM(sONE,sPROD(sC3,sSUM(sONE,sT1)))
117:       sCDFO=sUPPER(sSUM(sT2,sPROD(sT3,sT4)))
118:
119:
120:       RETURN
121:       END
```

FDREDUCE

```
   1:        SUBROUTINE FDREDUCE(vDRHS,vAO,vCO,vDO,sDAO,sDCO,sDDO
   2:       *                              ,vxDJ,sGDJ,sHDJ)
   3:        INCLUDE BIGN.STUFF
   4:        COMMON/DOMAIN/sRHOGO,sRHOGAMMA,sXOh,sCENTh,sRHOh,sCENTG,sRHOG
   5:       *             ,sRHOhB,sRHOPhB,sXOhB,sPHIO,sxKAPPA,sKAPPA
   6:       *             ,sRHOP,sRHO1
   7:        COMMON/REDUCE/sGO3,sGO4,sGO5,sGA3,sGA33,sGA5,sGA6
   8:        DIMENSION vDRHS(1),vAO(1),vCO(1),vDO(1),vxDJ(1)
   9:        DIMENSION vDJ(NDEG)
  10:
  11:
  12:C  Calculation of irregular inhomogeneous part
  13:C  ===========================================
  14:
  15:C  Determine irregular coefficients of inhomogeneous part
  16:C  ------------------------------------------------------
  17:        sT1=sPROD(sRHOGAMMA,sRHOGAMMA)
  18:        sT2=DCMPLX(-rG(vDRHS),rG(vDRHS))
  19:        sRO=vDRHS(1)
  20:        sR2=sQUOT(sSUM(vDRHS(3),sT2),sT1)
  21:        sR3=sQUOT(sSUM(vDRHS(4),sT2),sPROD(sT1,sRHOGAMMA))
  22:
  23:C  Calculate coefficients of irregular part of solution
  24:C  ----------------------------------------------------
  25:        sDAO=sNEG(sPROD(sHALF,sQUOT(sRO,sGA3)))
  26:        sDCO=sQUOT(sDIFF(sR2,sPROD(sDAO,sDIFF(sPROD(sTHREE,sGA33)
  27:       *     ,sPROD(sTWO,sGA5)))),sPROD(sTWO,sGA3))
  28:        sDDO=sQUOT(sSUM(sR3,sPROD(sPROD(sDAO,sTWO),sGA6)),sGA3)
  29:
  30:
  31:C  Calculate regular inhomogeneous part
  32:C  ====================================
  33:
  34:C  Store vDRHS
  35:C  -----------
  36:        CALL FEQUAL(vDRHS,vDJ)
  37:
  38:C  Add terms with sDAO
  39:C  -------------------
  40:        IF(rMAXABS(sDAO).GT.rZERO)THEN
  41:          CALL FsMULADD(vAO,sNEG(sDAO),vDJ)
  42:        ENDIF
  43:
  44:C  Add term with sDCO
  45:C  ------------------
  46:        IF(rMAXABS(sDCO).GT.rZERO)THEN
  47:          CALL FsMULADD(vCO,sNEG(sPROD(sTWO,sDCO)),vDJ)
  48:        ENDIF
  49:
  50:C  Add term with sDDO
  51:C  ------------------
  52:        IF(rMAXABS(sDDO).GT.rZERO)THEN
  53:          CALL FsMULADD(vDO,sNEG(sDDO),vDJ)
  54:        ENDIF
  55:
  56:C  Normalize properly
  57:C  ------------------
  58:        vDJ(1)=sZERO
  59:        vDJ(2)=sZERO
  60:        vDJ(3)=sZERO
  61:        vDJ(4)=sZERO
  62:
  63:C  Split off central part; calculate sGDJ and sHDJ
  64:C  ===============================================
  65:
```

FDREDUCE

```
66:C   Choose central values of coefficients and calculate difference
67:C   --------------------------------------------------------------
68:        CALL FSETZERO(vxDJ)
69:        DO 1 I=1,N1
70:        vxDJ(I)=saCONST(aCENT(vDJ(I)))
71:        vDJ(I)=sDIFF(vDJ(I),vxDJ(I))
72:      1 CONTINUE
73:
74:C   Scale down
75:C   ----------
76:        CALL FsDILAT(vDJ,sRHO1,vDJ)
77:
78:C   Collect higher order bounds in sHDJ
79:C   -----------------------------------
80:        sTH=sCONST(rH(vDJ))
81:        sT1=sABS(vDJ(N1-2))
82:        sT2=sABS(vDJ(N1-1))
83:        sT3=sABS(vDJ(N1  ))
84:        sHDJ=sUPPER(sSUM(sSUM(sSUM(sT1,sT2),sT3),sTH))
85:
86:C   Collect bounds on difference in sGDJ
87:C   ------------------------------------
88:        sTG=sCONST(rG(vDJ))
89:        vDJ(N1-2)=sZERO
90:        vDJ(N1-1)=sZERO
91:        vDJ(N1  )=sZERO
92:        vDJ(NERR)=sZERO
93:        sGDJ=sUPPER(sSUM(sCONST(rL1NORM(vDJ))
94:      *      ,sPROD(sTG,sPOWER(sRHO1,3))))
95:
96:
97:        RETURN
98:        END
```

We start the discussion with FDREDUCE. It is essentially identical to the subroutine FREDUCE and we therefore refer to Section 5.7. for the details. The function δR is no longer a constant function as R was in FvKI but is of the form

$$\delta R(z) = \delta R_o + \delta R_2 z^2 + \delta R_3 z^3$$

(see (5.8.13)). To split off the singular function δF_{sing} (see Table 5.8.1) we use the equations (L24-27)

$$\delta a_o = -\frac{1}{2}\delta R_o/\gamma_3 \equiv \text{sDA0} , \qquad (5.8.73)$$

$$\delta c_o = \frac{1}{2\gamma_3}(\delta R_2 - \delta a_o(3\gamma_3^2 - 2\gamma_5)) \equiv \text{sDC0} , \qquad (5.8.74)$$

$$\delta d_o = \frac{1}{\gamma_3}(\delta R_3 + 2\delta a_o\gamma_6) \equiv \text{sDD0} , \qquad (5.8.75)$$

where $\delta R_o \equiv$ sR0, $\delta R_2 \equiv$ sR2, $\delta R_3 \equiv$ sR3 (L16-21) and where $\gamma_3 \equiv$ sGA3, $\gamma_3^2 \equiv$ sGA33, $\gamma_5 \equiv$ sGA5 and $\gamma_6 \equiv$ sGA6 have already been calculated in L49-57 of FvKI, and are the Taylor coefficients of the function Γ, of order 3, 5 and 6 respectively. Note that the coefficients δa_o, δc_o and δd_o are zero if $\delta R(z) = O(z^k)$, $k \geq 4$, which is the case if $\delta\Gamma(z) = O(z^k)$, $k \geq 7$ (see (5.8.13)). We therefore test in L40, L46 and L52 if we need to reduce the right hand side at all. We again split into polynomial, lower order and higher order (L67-94) $\delta J = \delta J_{P,thin} + \Delta\delta J_G + \Delta\delta J_H$ and bound (L87-94)

$$\sup_{z \in \delta} \left| \frac{\Delta \delta J_G}{(z/\rho^+)^4} \right| \leq \left| \Delta \delta J_G^u \left(\frac{\rho^+}{\rho_\Gamma} \cdot \right) \right|_1 \quad (\equiv \text{SGDJ}) \equiv \text{SGDJ2} \tag{5.8.76}$$

and   (L79-84)

$$\sup_{z \in \delta} \left| \frac{\Delta \delta J_H}{(z/\rho^+)^{N-3}} \right| \leq \left| \Delta \delta J_H^u \left(\frac{\rho^+}{\rho_\Gamma} \cdot \right) \right|_1 \quad (\equiv \text{SHDJ}) \equiv \text{SGDJ2} . \tag{5.8.77}$$

These bounds correspond to the bounds $\Delta \delta J_{2,G}$ and $\Delta \delta J_{2,H}$ in Table 5.8.1, and the subroutine FDREDUCE is called with these arguments in CDREDUCE (L20) and we continue our discussion at that point.

The power series $\delta \hat{J}_1(x) \equiv \text{xDJ1}$ is given as an input to the subroutine CDREDUCE furthermore we are given the function $\delta J_{2,thin}^u(z) \equiv \text{vxDJ2}$ as an output from FDREDUCE. We calculate the power series $\delta \hat{J}_2(x) \equiv \text{xDJ2}$ in L28-31 (and take care to adapt the degree, if necessary, in L35-37, L46), and sum the two contributions into $\delta \hat{J}(x) = \delta \hat{J}_1(x) + \delta \hat{J}_2(x) \equiv \text{xDJ}$. We are also given as an input to CDREDUCE the general and higher order bounds $\Delta \delta J_{1,G} \equiv \text{SGDJ1}$ and $\Delta \delta J_{1,H} \equiv \text{SHDJ1}$ and as an output from FDREDUCE the bounds $\Delta \delta J_{2,G} \equiv \text{SGDJ2}$ and $\Delta \delta J_{2,H} \equiv \text{SHDJ2}$ we add these contributions in L52-53 to yield

$\Delta \delta J_G = \Delta \delta J_{1,G} + \Delta \delta J_{2,G} \equiv \text{SGDJ}$ and to $\Delta \delta J_H = \Delta \delta J_{1,H} + \Delta \delta J_{1,H} \equiv \text{SHDJ}.$
So the only thing which remains to be done is to establish bounds of the form (5.7.20), (5.7.21) for the Borel transform $\delta j_B(t)$ of the function $\delta j(z) = \delta J((\frac{\kappa}{4}z)^{1/2})$, the normal form of the function $\delta J(z) = \delta J_1(z) + \delta J_{2,thin}(z)$. This is done in L59-104 which we shall explain next, but we note already here that once

we have established these bounds we can use (5.7.26) and (5.7.27) to establish, L108-117, bounds on the Borel transform $\delta f_{B,\mathcal{L}}(t)$ or $\delta f_{\mathcal{L}}(z) = \delta F(z^{1/2})$ (see Table 5.8.1 for the definition of δF) and we can apply afterwards the sequence of subroutines CxFØ, FSUMBOREL, FADDGENERAL and FCOMPLETE to construct the vector $\delta h_B^u(z) \equiv vDhB$ (L72, L81, L85-86 and L90-93 of FvDKI, respectively).

A part of the quantities which we need in order to explain L59-104 have already been calculated in the preparatory step in L_M212-239. We proceed as follows : Let $\delta f(z)$ be the normal form of $\delta F(z)$ and $\delta f_B(t)$ its Borel transform. We then need constants $C_{\delta f}$, $E_{\delta f} > 0$, such that

$$|\delta f_B(t)| \leq C_{\delta f} \exp(E_{\delta f}|t|) \tag{5.8.78}$$

for $t \in S_{2,\beta}$. Since the function δj (the normal form of the in-homogeneity δJ) is the sum of two functions δ_{j_1} and δ_{j_2} (the normal forms of the functions δJ_1 and δJ_2) we shall first establish bounds for δj_1 and δj_2 separately and combine these to prove that for $t \in S_{2,\beta}$,

$$|\delta j_B(t)| \leq |t| C_{\delta j} \exp(E_{\delta j}|t|) . \tag{5.8.79}$$

We can then apply Ecalle's bounds (5.7.26), (5.7.27) to prove (5.8.78). Note that δJ_2 is analytic not only on δ but even on $D(0,\rho_\Gamma)$, and we could therefore bound $\delta j_{2,B}(t)$ by $|\delta j_{2,B}(t)| \leq |t| C_{\delta j_2} \exp(E_{\delta j_2}|t|)$ with $E_{\delta j_2} = (\frac{\kappa}{4\rho_\Gamma^2})$. However it turns out that the $E_{\delta \hat{\jmath}_1}$, the corresponding bound for $\delta \hat{\jmath}_{1,B}(t)$, can not be chosen as small as $E_{\delta \hat{\jmath}_2}$, and in view of the fact that

$$C_{\delta \hat{\jmath}} = C_{\delta \hat{\jmath}_1} + C_{\delta \hat{\jmath}_2} ,$$

(5.8.80)

$$E_{\delta \hat{\jmath}} = \max (E_{\delta \hat{\jmath}_1} , E_{\delta \hat{\jmath}_2})$$

we have decided to first find $C_{\delta \hat{\jmath}_1}$ and $E_{\delta \hat{\jmath}_1}$ and to choose then $E_{\delta \hat{\jmath}_2} = E_{\delta \hat{\jmath}_1}$ in order to optimize the bound for $C_{\delta \hat{\jmath}_2}$. By defini-tion we have

$$\delta \jmath_1(z) = 2 \cdot \{ (f'(z) \cdot z^2 + j'(z) \cdot z^2) \} \{ \frac{1}{z} \frac{(\frac{1}{z}g(z))^{1/2}}{g'(z)} \} \{ \frac{1}{z}\delta g(z))^{1/2} \}$$

(5.8.81)

where $\delta \jmath_1$, f, j, q and δg are the normal forms of the func-tions δJ_1, F, $J_{P,thin}$, Γ and $\delta \Gamma$, respectively. Define now the function $\Delta(z)$ by the equation

$$\Delta(z) = (\frac{1}{z}\delta g(z))^{1/2} ,$$

(5.8.82)

and the function $q(z)$ by

$$q(z) = \frac{1}{z} \frac{\Delta(z)}{g'(z)} - \frac{1}{z} - \frac{3}{2} ,$$

(5.8.83)

then we get, if we take the Borel transform of the equation (5.8.81)

$$\delta \jmath_{1,B}(t) = 2\{ t \cdot f_B(t) + t j_B(t) \} * \{ \delta'(t) + \frac{3}{2}\delta(t) + q_B(t) \}$$

$$* \{ \Delta_B(t) \} ,$$

(5.8.84)

where $\delta(t)$ is the Dirac δ-distribution. We have thus

$$\delta j_{1,B}(t) = 2 \cdot \{t \cdot f_B(t) + t \cdot j_B(t)\} * \{\Delta_B(0)\delta(t) + \Delta'_B(t)$$

$$+ \frac{3}{2}\Delta_B(t) + \int_o^t q_B(t-t_1)\Delta_B(t_1)dt_1\} \ . \tag{5.8.85}$$

We shall bound below, for $t \in S_{2,\beta}$

$$|q_B(t)| \leq C_q \exp(E_q|t|) \ , \tag{5.8.86}$$

$$|\Delta_B(t)| \leq C_\Delta \exp(E_\Delta|t|) \ , \tag{5.8.87}$$

$$|\Delta'_B(t)| \leq C_{\Delta'} \exp(E_{\Delta'}|t|) \ , \tag{5.8.88}$$

and hence, we get

$$\left| \int_o^t q_B(t-\tau)\Delta_B(\tau)d\tau \right| \leq C_q C_\Delta \exp(\max\{E_q,E_\Delta\}|t|) \cdot |t| \ .$$

For $\delta j_{1,B}(t)$ we then find, using

$$|t|^k < \exp(\frac{k}{e}|t|) \ ,$$

$$|\delta j_{1,B}(t)| \leq 2 \cdot |t| \cdot ((C_f + C_j)\Delta_B(0) + (C_{\Delta'} + \frac{3}{2}C_\Delta)(C_f + C_j)\frac{1}{2}$$

$$+ C_q \cdot C_\Delta \cdot (C_f + C_j) \cdot \frac{1}{6}) \cdot \exp(\max(E_f,E_j)$$

$$+ \max(\max(E_\Delta,E_{\Delta'}) + \frac{2}{e} \ , \ E_q + E_\Delta + \frac{3}{e})) \ . \tag{5.8.89}$$

The constants C_f and E_f have been calculated in Section 5.7, (5.7.26). Furthermore we replace the bound (5.7.19a) for the function j_B by

$$|j_B(t)| \leq C_j \exp(E_j |t|) \, . \tag{5.8.90}$$

C_j , $E_j > 0$ will be given below. From Section 5.7 we have that E_j could be choosen smaller than E_f , but in order to optimize the bound (5.8.89) we choose $E_j = E_f$. Furthermore it is easy to see that in order to further optimize the bound (5.8.89) one has to choose $E_{\Delta'} > E_\Delta$, and even more explicitly

$$E_{\Delta'} = E_q + E_\Delta + \frac{1}{e} \, . \tag{5.8.91}$$

This leads to

$$|\delta j_{1,B}(t)| \leq |t| \cdot C_{\delta \hat{j}_1} \exp(E_{\delta \hat{j}_1} |t|) \, , \tag{5.8.92}$$

where

$$C_{\delta \hat{j}_1} = 2((C_f + C_j) \Delta_B(0) + (C_{\Delta'} + \tfrac{3}{2} C_\Delta)(C_f + C_j) \tfrac{1}{2}$$

$$+ C_q \cdot C_\Delta \cdot (C_f + C_j) \tfrac{1}{6}) \, , \tag{5.8.93}$$

$$E_{\delta \hat{j}_1} = E_f + E_q + E_\Delta + \frac{3}{e} \, . \tag{5.8.94}$$

This completes the discussion of the bound for δj_1 and we proceed now to give explicit bounds for the constants which are involved.

We have $(L_M 231-239)$

$$E_j = E_f \ (\equiv sEF\emptyset) \equiv sEJ \ , \tag{5.8.95}$$

$$C_j \leq E_j \left| J^u (\frac{\rho}{\rho_\Gamma} .) \right|_1 \equiv sCJ \ , \tag{5.8.96}$$

where $\rho = (\frac{\kappa}{4E_j})^{1/2}$. We next bound C_q and E_q . We fix $(L_M 212)$
$\rho = 0.45$ and define $(L_M 214)$,

$$E_q = \frac{\kappa_\rho}{4\rho^2} \equiv sEQ \ . \tag{5.8.97}$$

We then have $(L_M 216-219)$

$$C_q \leq (E_q)^2 \cdot \left| \left(\frac{1}{G_o^{u'} (G_o^u (\frac{\rho}{\rho_o} z))} \right) \cdot \left(\frac{1}{G_o^{u'} (\frac{\rho}{\rho_o} z)} \right) - 1 - \frac{3}{2} \frac{4}{\kappa} \rho^2 z^2 \right|_1$$

$$\equiv sCQ \ . \tag{5.8.98}$$

Our choice for ρ is dictated by the fact that we need the inverse of G_o' to exist in (5.8.98). The other bounds of (5.8.85)-(5.8.88) depend on $\delta\Gamma$ and are calculated in CDREDUCE. We choose E_Δ as small as possible, namely (L62-63)

$$E_\Delta = \frac{\kappa_\rho}{4\rho_\Gamma^2} \equiv sEDEL \ , \tag{5.8.99}$$

and get (L64-65)

$$C_\Delta \leq E_\Delta \left| \delta\Gamma^u (.) \right|_1 \equiv sCDEL \ . \tag{5.8.100}$$

Next we choose $E_{\Delta'}$ as described in (5.8.91), namely (L67)

$$E_{\Delta'} = E_q + E_\Delta - \frac{1}{e} \equiv \text{sEDELP} , \qquad (5.8.101)$$

choose then $\rho = (\frac{\kappa_p}{4E_{\Delta'}})^{1/2}$ (L68), and get

$$C_{\Delta'} \leq |(E_{\Delta'})^2 (\delta\Gamma^u(z)/z^3) - (\delta\Gamma^u/z^3)|_{z=0}|_1 \equiv \text{sCDELP} . \qquad (5.8.102)$$

Using these constants and furthermore that $\Delta_B(0) = \sqrt{\frac{4}{\kappa}}(\frac{1}{z^3}\delta\Gamma(z))|_{z=0}$

we can calculate the constants $C_{\delta\hat{\jmath}_1} \equiv \text{sCDJ1H}$ (L77-79) and

$E_{\delta\hat{\jmath}_1} \equiv \text{sEDJ1H}$ (L74-75). For the bound on $\delta\hat{\jmath}_2$ we proceed as des-

cribed above, we choose (L83)

$$E_{\delta\hat{\jmath}_2} = E_{\delta\hat{\jmath}_1} \quad (\equiv \text{sEDJ1H}) \equiv \text{sEDJ2H} , \qquad (5.8.103)$$

put $\rho = (\kappa_p/4E_{\delta\hat{\jmath}_2})$ (L84) and get (L85-87)

$$C_{\delta\hat{\jmath}_2} \leq (E_{\delta\hat{\jmath}_2})^2 |\delta J^u_{2,\text{thin}}(\frac{\rho}{\rho_\Gamma}\cdot)|_1 . \qquad (5.8.104)$$

We add now the bounds (5.8.93) and (5.8.104) for $\delta\hat{\jmath}_{1,B}(t)$ and

$\delta\hat{\jmath}_{2,B}(t)$ respectively and get so the bound (5.8.79) for $\delta\hat{\jmath}_B$. We

have (L91-92) $E_{\delta\hat{\jmath}} = E_{\delta\hat{\jmath}_1} = E_{\delta\hat{\jmath}_2}$ ($\equiv$ sEDJ1H) $\equiv$ sEDJH, and

$C_{\delta\hat{\jmath}} = C_{\delta\hat{\jmath}_1} + C_{\delta\hat{\jmath}_2} \equiv \text{sCDJH}$. In order to optimize the bound for

$C_{\delta f}$ we reestimate the constant C_γ. We put (L96)

$$E_\gamma = E_{\delta\hat{\jmath}} \equiv \text{sEI2ETA} , \qquad (5.8.105)$$

then $\rho = (\kappa_p/4E_\gamma)^{1/2}$ (L97) and get (L98-104)

$$C_\gamma \leq (E_\gamma)^2 \left| \frac{1}{(H^4(\frac{\rho}{\rho_\Gamma}z))^2} - 1 - \frac{4}{\kappa}\rho^2 z^2 \right|_1 .$$

(5.7.26) is now readily applied and we get (L108-112)

$$E_{\delta f} = \max(E_{\delta j} + 1, \; E_{\delta j} + C_\gamma(C_3 + 1)) \equiv \text{sEDF}\emptyset , \qquad (5.8.106)$$

furthermore (L114-117)

$$C_{\delta f} = C_3 C_{\delta j} + \{C_3 C_{\delta j}/C_\gamma (1+C_3)^2\}(1 + C_3(1 + C_\gamma(1 + C_3)))$$

$$\equiv \text{scDF}\emptyset . \qquad (5.8.107)$$

This completes our discussion of the subroutines FDREDUCE and
CDREDUCE. By the remark above we proceed now as in the case of
the subroutine FvKI, i.e. we apply one after the other the subroutine CxF∅, FSUMBOREL, FADDGENERAL and FCOMPLETE (see subroutine FvDKI) and we end up with $\delta h_B^u \equiv$ vDhB being defined
(L93 of FvDKI). We continue our discussion of FvDKI there.
We follow now exactly the equations defining DK (see Section 5.3).
L95-193 of FvDKI are then selfexplaining if one notes that all
the remaining quantities which are independent of $\delta\Gamma$ are calculated in $L_M 440$-497 : We have $L_M 445$-448

$$G_O^{u'}(\frac{1}{\rho_\Gamma}T_{-A}(-x_{oh} + \rho_h z)) \equiv \text{vDZG}\emptyset\emptyset\text{ARG} ,$$

$$G_O^{u'}((G_O^u)^i \circ (\frac{1}{\rho_\Gamma}T_{-A}(-x_{oh} + \rho_h z))) \equiv \text{vDZG}\emptyset\text{ARG}[\; ,i] ,$$

$$i = 1,\ldots,\text{IBOOT}$$

and $(L_M 452-453)$

$$\frac{\rho_o}{\rho_{h_B}} G_o^{u'} ((G_o^u)^{IBOOT+1} \circ (\frac{1}{\rho_\Gamma} T_{-A} (-x_{oh} + \rho_h z))) \equiv vDZhLASTG\emptyset$$

Furthermore we have $(L_M 457-466)$

$$h^{u'} (z) \equiv vDZh .$$

(See (5.3.4) for the definition of the function $h^u(z)$.)   Next we calculate $(L_M 470-477)$

$$\delta_A h^u (z) \equiv vDAh ,$$

and $(L_M 481-487)$

$$\delta_\tau \widetilde{G}^u (z) \equiv vDTAUG .$$

(See (5.3.7) for the definition of the function $\widetilde{G}^u(z)$.)   Finally we have $(L_M 491-497)$

$$\delta_A \widetilde{G}_o^u (z) \equiv vDAIG\emptyset I .$$

(See (5.3.9) for the definition of the function $\widetilde{G}_o^u(z)$.)

   This completes our discussion of the subroutine FvDKI. DK is defined. Running the program proves our main result of this paper which is Theorem 5.4.3.

This proves that the equations (1.7)-(1.10) have a solution. In Section 6 we next show how, starting from these equations, solutions to Feigenbaum's equation can be constructed for N large but finite. That part is proved <u>without</u> the help of a computer.

6. The Equations for $N < \infty$

6.1. Outline of the Method and Results

In this section we outline our strategy for solving (1.3)-
(1.6). Note again that in this part we need no longer a computer
to prove our theorems. The analysis is based however on the results obtained in Section 5.

To adapt the notation to the one used in (1.7)-(1.10) we rewrite the system (1.3)-(1.6) as follows :

$$f_N(z) = \frac{1}{\lambda_N} f_N(g_N(z)) , \qquad\qquad (6.1.1)$$

$$f_N(0) = \beta_N , \qquad\qquad (6.1.2)$$

$$g_N(z) = \alpha_N (f_N(|\lambda_N|^{2N} z))^{2N} . \qquad\qquad (6.1.3)$$

We want to solve these equations for $\alpha_N = 1$ and $\beta_N = 1$, but
again, since α_N and β_N only fix the normalization of the solution we shall feel free to omit these conditions in the sequel,
replacing them by α_N , $\beta_N \neq 0$.

The key idea for solving (6.1.1)-(6.1.3) for large N is
to consider λ_N as the independent variable instead of N, i.e.
to fix λ_N near to -1 and to search functions f_N and g_N and
a number N, solving (6.1.1)-(6.1.3) for this particular value of
λ_N. To adapt the notation to this idea we fix a small $\varepsilon > 0$,

and define

$$\lambda_N \equiv \lambda_\varepsilon = -(1 - 2\varepsilon)^{1/2} . \qquad (6.1.4)$$

Furthermore, we consider τ_ε instead of N to be the quantity which needs to be determined, where τ_ε is defined through

$$\tau_\varepsilon = |\lambda_\varepsilon|^{2N} . \qquad (6.1.5)$$

In terms of these definitions we get for (6.1.1)-(6.1.3) the following equivalent system :

$$f_\varepsilon(z) = \frac{1}{\lambda_\varepsilon} \cdot f_\varepsilon(g_\varepsilon(z)) , \qquad (6.1.6)$$

$$f_\varepsilon(0) = 1 , \qquad (6.1.7)$$

$$g_\varepsilon(z) = \alpha_\varepsilon \cdot \exp\left(\frac{\log \tau_\varepsilon}{\log|\lambda_\varepsilon|} \cdot \log(f_\varepsilon(\tau_\varepsilon z))\right) , \qquad (6.1.8)$$

$$g'_\varepsilon(x_\varepsilon) = \lambda_\varepsilon , \quad g_\varepsilon(x_\varepsilon) = x_\varepsilon . \qquad (6.1.9)$$

The strategy for solving (6.1.6)-(6.1.9) is exactly the same as that for solving (1.7)-(1.10). We formulate the task to solve the system as a fixed point problem. We follow the road which we have paved in order to define the operator K and we are led to define the operator $K_\varepsilon : \{A, G_0\} \to \{\tilde{A}, \tilde{G}_0\}$ by the following sequence of steps.

i) given $\{A, G_0\} \in A_1$, define

$$G_{o,\varepsilon} = |\lambda_\varepsilon| \cdot G_o \ , \qquad (6.1.10)$$

ii) show that the equation

$$F_{o,\varepsilon}(z) = \frac{1}{\lambda_\varepsilon} F_{o,\varepsilon}(G_{o,\varepsilon}(z)) \ , \qquad (6.1.11)$$

has a solution on $\mathcal{D}_\varepsilon$ (see next section)

iii) define

$$h_\varepsilon^u(z) = \frac{F_{o,\varepsilon}(T_{-A}(-x_{oh}+\alpha_h z))}{F_{o,\varepsilon}(T_{-A}(-x_{oh}))} - 1 \ , \qquad (6.1.12)$$

iv) solve the following equation for τ_ε

$$\frac{x_{oo}\tau_\varepsilon}{\rho_h} \cdot \frac{\log\tau_\varepsilon}{\log|\lambda_\varepsilon|} \cdot \frac{h_\varepsilon^{u\,'}\left(\dfrac{\tau_\varepsilon x_{oo}-\alpha_h}{\rho_h}\right)}{1+h_\varepsilon^u\left(\dfrac{\tau_\varepsilon x_{oo}-\alpha_h}{\rho_h}\right)} = \lambda_\varepsilon \ , \qquad (6.1.13)$$

v) set

$$\alpha_\varepsilon = x_{oo} \cdot \exp\left(\frac{\log\tau_\varepsilon}{\log|\lambda_\varepsilon|} \cdot \log\left(1+h_\varepsilon^u\left(\frac{\tau_\varepsilon\rho_G}{\rho_h}z + \frac{\tau_\varepsilon(\alpha_G+x_{oo})-\alpha_h}{\rho_h}\right)\right)\right) \ , $$
$$\qquad (6.1.14)$$

vi) define

$$\tilde{G}_\varepsilon^u(z) = \frac{\alpha_\varepsilon}{\rho_G} \cdot \exp\left(\frac{\log\tau_\varepsilon}{\log|\lambda_\varepsilon|} \cdot \log\left(1 + \right.\right.$$

$$h_\varepsilon^u\left(\frac{\tau_\varepsilon\rho_G}{\rho_h}z + \frac{\tau_\varepsilon(\alpha_G+x_{oo})-\alpha_h}{\rho_h}\right)\right) - \frac{x_{oo}+\alpha_G}{\rho_G} \ , \qquad (6.1.15)$$

vii) determine

$$A_\varepsilon = -\frac{1}{4}[\frac{\lambda_\varepsilon^2}{x_{oo}} + x_{oo} \frac{\log\tau_\varepsilon}{\log|\lambda_\varepsilon|}(\frac{\tau_\varepsilon}{\rho_h})^2 \cdot$$

$$(\frac{h_\varepsilon^{u"}(\frac{\tau_\varepsilon x_{oo}^{-\alpha_h}}{\rho_h})}{1+h_\varepsilon^u(\frac{\tau_\varepsilon x_{oo}^{-\alpha_h}}{\rho_h})} - (\frac{h_\varepsilon^{u'}(\frac{\tau_\varepsilon x_{oo}^{-\alpha_h}}{\rho_h})}{1+h_\varepsilon^u(\frac{\tau_\varepsilon x_{oo}^{-\alpha_h}}{\rho_h})})^2)] , \qquad (6.1.16)$$

viii) define

$$G_{o,\varepsilon}^u(z) = \frac{1}{\rho_o}T_{-\tilde{A}_\varepsilon}(\alpha_G + \rho_G\tilde{G}_\varepsilon^u(\frac{1}{\rho_G}(T_{\tilde{A}_\varepsilon}(\rho_o z) - \alpha_G))) \qquad (6.1.17)$$

ix) define

$$\tilde{A} = \tilde{A}_\varepsilon , \quad \tilde{G}_o = \frac{1}{|\lambda_\varepsilon|}\tilde{G}_{o,\varepsilon} \qquad (6.1.18)$$

x) define

$$K_\varepsilon(\{A,G_o\}) = \{\tilde{A},\tilde{G}_o\} . \qquad (6.1.19)$$

<u>Lemma 6.1.1. If</u> K_ε <u>is defined and has a fixed point, then (6.1.6)-</u>
<u>(6.1.9) has a solution.</u>

This is obvious from the definition of K_ε.

We now proceed as follows.

Let $\{A^*,G_o^*\} \in D_K$ be the fixed point of the operator K, as constructed in Section 5. We give ourselves a neighborhood E of $\{A^*, G_o^*\}$ in A_1 as follows :

$$E = \{\{A,G_o\} \in A_1 \mid \| \{A,G_o\} - \{A^*,G_o^*\} \| < \theta\} \ , \qquad (6.1.20)$$

where θ is a strictly positive number (independent of ε) but small enough such that we have the inclusion $E \subset D_K$.

It follows immediately from this definition, that the operator K is defined on E, maps E into itself and that the tangent map $DK_{\{A,G_o\}}$ is bounded in norm by the bound 2) given in Theorem 5.4.3, uniformly in $\{A,G_o\} \in E$.

Our results concerning the operator K_ε are collected in the following two theorems.

Theorem 6.1.2. The operator K_ε is defined on E. We have the bound

$$\| K_\varepsilon(\{A,G_o\}) - K(\{A,G_o\}) \| \le \theta_1(\varepsilon) \ , \qquad (6.1.21)$$

where the bound $\theta_1(\varepsilon)$ is uniform for $\{A,G_o\} \in E$, continuous as a function of ε and where $\lim_{\varepsilon \to 0} \theta_1(\varepsilon) = 0$.

It will be shown that K_ε is once continuously differentiable as a map from E into A_1 and that the derivative DK_ε is a compact operator on A_o. In fact, by using the analyticity properties it can be shown that K_ε is infinitely often differentiable. We do not present the formulas describing DK_ε here. They are obtained in complete analogy to the case of the operator DK by linearizing the formulas describing K_ε.

Theorem 6.1.3. We have the bound

$$\|DK_{\varepsilon,\{A,G_o\}} - DK_{\{A,G_o\}}\| \leq \theta_2(\varepsilon) , \qquad (6.1.22)$$

where the bound $\theta_2(\varepsilon)$ is uniform for $\{A,G_o\} \in E$, continuous as a function of ε and where $\lim_{\varepsilon \to 0} \theta_2(\varepsilon) = 0.1220127 \cdot 10^{-9}$.

For the proof of these two theorems see the following sections.

The main result of this book is the following theorem.

Theorem 6.1.4. The operator K_ε has a fixed point $\{A^{*,\varepsilon}, G_o^{*,\varepsilon}\}$ in $E \subset D_K \subset A_1$. It is unique in E. We have the bound

$$\|\{A^{*,\varepsilon}, G_o^{*,\varepsilon}\} - \{A^*, G_o^*\}\| \leq \frac{1}{1-\rho}\theta_1(\varepsilon) , \qquad (6.1.23)$$

where $\rho =$ is the bound on the tangent map DK given in Theorem 5.4.3.

The existence and uniqueness of a fixed point follows again from the contraction mapping principle, using the bound

$$\theta_1(\varepsilon) \leq (1 - (\rho + \theta_2(\varepsilon))) \cdot \theta ,$$

which is satisfied if ε is small enough. The bound (6.1.23) follows using the inequalities

$$\| \{A^{*,\varepsilon}, G_o^{*,\varepsilon}\} - \{A^*, G_o^*\} \| \leq \| K_\varepsilon (\{A^{*,\varepsilon}, G_o^{*,\varepsilon}\}) - K(\{A^{*,\varepsilon}, G_o^{*,\varepsilon}\}) \|$$

$$+ \| K(\{A^{*,\varepsilon}, G_o^{*,\varepsilon}\}) - K(\{A^*, G_o^*\}) \|$$

$$\leq \theta_1(\varepsilon) + \rho \| \{A^{*,\varepsilon}, G_o^{*,\varepsilon}\} - \{A^*, G_o^*\} \| .$$

Theorem 6.1.4 provides the proof of the existence and local uniqueness of a solution of the equation (2) for N large.

6.2. Outline of the proof of Theorems 6.1.2 and 6.1.3

In this section we give an outline of the proof of Theorem 6.1.2 and Theorem 6.1.3. Since these proofs are rather long and since the constructions are similar in both cases we elaborate only on the equations for Theorem 6.1.2, which are somewhat easier to handle.

The idea is to discuss step by step the equations defining the operator K_ε and to bound step by step the difference between the objects appearing in these equations and objects which we explicitly construct, using the results of Section 5 as an input.

As an example of this technique we first prove in this section the following theorem.

Theorem 6.2.1. Let $\{A, G_o\} \in E$, and let h^u be the function defined through (5.3.4), and corresponding to G_o. There is an $\varepsilon_o > 0$ such that for $\varepsilon < \varepsilon_o$ the functions h_ε^u defined through (6.1.12), and corresponding to G_o, are analytic on the unit disc and satisfy the bound

$$\| h_\varepsilon^u(.) - \varepsilon h^u(.) \|_1 < C \cdot \varepsilon^2 \cdot (\log \varepsilon)^2 , \tag{6.2.1}$$

where $\| . \|_1$ denotes the ℓ^1-norm of the Taylor coefficients when the functions are expanded at $z = 0$, and where C is some positive constant independent of ε.

Proof of the Theorem 6.2.1 : The main problem (and one of the main issues of this book) is to construct the solution of (6.1.11) on the domain $\mathcal{D}_\varepsilon$ which we now describe.

Let $S^\pm$ be the sectors which are defined in (5.2.5) and let

$$S = \{ z^2 \,|\, z \in S^+ \} \equiv \{ z^2 \,|\, z \in S^- \} . \tag{6.2.2}$$

For every $\varepsilon > 0$ we define the disc

$$d_\varepsilon = \{ z \in \mathbb{C} \,\big|\, |z| < \tfrac{2}{5}\varepsilon^{1/2} \} .$$

Next we fix $\varepsilon_o > 0$ (very) small and truncate our sectors $S^\pm$ (see (5.2.5)) by a tiny bit, i.e. we define

$$S_T^\pm = \{ z \in S^\pm \,\big|\, |z| < \rho_\Gamma - \Delta \} \tag{6.2.3}$$

where Δ is chosen small enough but fixed, such that $z^2 + \kappa\varepsilon\in S$, for every $\{A,G_o\}\in E$, $0 < \varepsilon < \varepsilon_o$ and $z\in S_T^+ \cup S_T^-$.

Using the bound for κ given in Lemma 5.4.4 and the above definition of S, one also easily checks that in addition $z^2 + \kappa\cdot\varepsilon\in S$ for every $\{A,G_o\}\in E$, $\varepsilon < \varepsilon_o$, and $z\in d_\varepsilon$.

We define now the domain $\mathcal{D}_\varepsilon$ by the equation

$$\mathcal{D}_\varepsilon = S_T^+ \cup S_T^- \cup d_\varepsilon \ . \tag{6.2.4}$$

For convenience later on we use the notation

$$W_\varepsilon = \mathcal{D}_\varepsilon \backslash d_\varepsilon \tag{6.2.5}$$

for the wings of our butterfly-shaped domain.

Proposition 6.2.2. The domain $\mathcal{D}_\varepsilon$ is invariant, i.e. for every $\{A,G_o\}\in E$ and $G_{o,\varepsilon}$ as defined through (6.1.10) we have

$$G_{o,\varepsilon}(\mathcal{D}_\varepsilon) \subset \mathcal{D}_\varepsilon \ .$$

Proof : For every $\{A,G_o\}\in E$ we have by Lemma 5.7.2 that $G_o(S^\pm) \subset S^\mp$. Furthermore it is clear that $|\lambda_\varepsilon|S^\pm \subset S^\pm$, and therefore $G_{o,\varepsilon}(S^\pm) \subset S^\mp$. It remains to be shown that $G_{o,\varepsilon}(d_\varepsilon) \subset d_\varepsilon$, this however follows readily using the definition of d_ε and the form of the Taylor expansion of $G_{o,\varepsilon}$ at 0 and $|\lambda_\varepsilon| < 1$ (see (6.1.4)). This proves the proposition.

We now elaborate on (6.1.11). Consider $\{A,G_o\}\in E$. We have

seen in Section 5.7 and Section 2.5 that there are numbers a_o,

c_o, d_o and functions F_1 and F_2 defined by $\{A,G_o\}$. Closer

inspection shows that these objects do not depend on A. We have

thus a map from G, the second component of E (i.e. the set of

G_o's we obtain by varying $\{A,G_o\}$ in E), to a_o, c_o, d_o, F_1

and F_2.

We define now a function $f_{o,\varepsilon}$ which will turn out to be a

good approximate solution to the equation

$$F_{o,\varepsilon}(z) = \frac{1}{\lambda_\varepsilon} \cdot F_{o,\varepsilon}(|\lambda_\varepsilon| G_o(z)) ,$$

i.e. to (6.1.11).

Definition 6.2.3. Let $G_o \in G$ and let a_o, c_o, d_o be the numbers

and F_1 and F_2 be the functions constructed in Section 5.7 and

Section 2.5 and corresponding to G_o. We define the function

$$F^o_{o,\varepsilon}(z) = \frac{z}{\sqrt{z^2+\kappa\varepsilon}} \cdot F_{1,\varepsilon}(z^2) + \varepsilon \cdot \frac{z^4}{\sqrt{z^2+\kappa\varepsilon}^3} \cdot F_{2,\varepsilon}(z^2) ,$$

where

$$F_{1,\varepsilon} = 1 + \varepsilon \cdot c_o \log(1+\frac{z^2}{\kappa\varepsilon}) - \varepsilon \cdot c_o \frac{z^2}{z^2+\kappa\varepsilon} + \varepsilon F_1^\varepsilon(z^2) ,$$

$$F_{2,\varepsilon} = d_o + F_2^\varepsilon(z^2) ,$$

where

$$F_1^\epsilon(z^2) = z^2 \cdot F_1(z^2 + \kappa\epsilon) \ ,$$

$$F_2^\epsilon(z^2) = z^2 \cdot F_2(z^2 + \kappa\epsilon) \ ,$$

and where $\kappa = 6/G_0'''(0)$ $(\equiv -2a_0$ by construction).

Remark : The ansatz for $F_{0,\epsilon}^0$ may seem mysterious to the reader. There is however a systematic way to produce it : a study of the order of magnitude of the coefficients in a power series ansatz for $F_{0,\epsilon}^0$ suggests to represent $F_{0,\epsilon}^0$ as $F_{0,\epsilon}^0 = x \cdot f_{1,\epsilon}(x^2/\epsilon)$ $+ x^2 \cdot f_{2,\epsilon}(x^2/\epsilon)$, to choose $z = x^2/\epsilon$ as the new variable and to use a power series expansion for $f_{1,\epsilon}$ and $f_{2,\epsilon}$ in powers of ϵ, i.e. $f_{1,\epsilon}(z) = \sum\limits_{i=0}^{\infty} f_{1,i}(z) \cdot \epsilon^i$, $f_{2,\epsilon}(z) = \sum\limits_{i=0}^{\infty} f_{2,i}(z) \cdot \epsilon^i$. In these variables $G_{0,\epsilon}$ is of the form $\lambda_\epsilon \cdot \epsilon^{1/2} z^{1/2} H_0(\epsilon^{1/2} z^{1/2})$. One develops then formally in powers of $\epsilon^{1/2}$ and is lead to a recursive system of equations, with a first order differential operator acting on $f_{1,i}$. This system can be solved order by order. Keeping terms which are globally (i.e. on $\mathcal{D}_\epsilon$) of the same order of magnitude leads to the ansatz given in Definition 6.2.3. A systematic improvement of such an ansatz is possible.

In the next lemma we shall bound the error which we make when we use $F_{0,\epsilon}^0$ as an approximate solution to (6.1.11). But to formulate the result we first need some notation.

i) Let $G_0 \in G$. By definition, G_0 is of the form $G_0(z) = -z \cdot H_0(z)$, and $H_0(z) = 1 - \frac{1}{\kappa}z^2 + \mathcal{O}(z^3)$. We reserve the symbol H_0 for functions obtained that way.

ii) Let $G_0 \in G$, and $G_{0,\epsilon}$ defined through (6.1.10), i.e.

$G_{o,\varepsilon}(z) = |\lambda_\varepsilon| \cdot G_o(z) = \lambda_\varepsilon \cdot z \cdot H_o(z)$. The set of $G_{o,\varepsilon}$'s which we get that way we denote G_ε.

iii) For the rest of this section we shall use the symbol B_ε for a function satisfying

$$\sup_{0<\varepsilon<\varepsilon_o} \sup_{z\in D_\varepsilon} |B_\varepsilon(z)| \leq L .$$

This function changes meaning for each occurrence and L may depend on the realization of B_ε (as the order symbol). The symbol $\mathcal{L}_\varepsilon$ will be used for the analogous local bound, i.e. for a function satisfying

$$\sup_{0<\varepsilon<\varepsilon_o} \sup_{z\in d_\varepsilon} |\mathcal{L}_\varepsilon(z)| < L ,$$

and the symbol W_ε for a function bounded on W_ε

$$\sup_{0<\varepsilon<\varepsilon_o} \sup_{z\in W_\varepsilon} |W_\varepsilon(z)| < L .$$

We can now formulate the following lemma.

<u>Lemma 6.2.4.</u> <u>Let for</u> $G_o \in G$, $G_{o,\varepsilon} \in G_\varepsilon$ <u>be given through (6.1.10) and</u> $F_{o,\varepsilon}^o$ <u>through Definition 6.2.3. The function</u>

$$r_{o,\varepsilon}^o(z) = \frac{1}{\lambda_\varepsilon} F_{o,\varepsilon}^o(G_{o,\varepsilon}(z)) - F_{o,\varepsilon}^o(z)$$

<u>is defined and analytic on</u> D_ε , <u>furthermore we have the bound</u>

$$r^o_{o,\varepsilon} = (\frac{z}{\sqrt{z^2+\kappa\varepsilon}})^3 \cdot \varepsilon^2 \cdot \log \varepsilon \cdot B_\varepsilon(z) \ .$$

<u>Proof of Lemma 6.2.4</u> : First remember that by definition of $\mathcal{D}_\varepsilon$,

$z^2 + \kappa\varepsilon \in S$ for every $z \in \mathcal{D}_\varepsilon$. Therefore $F_1(z^2 + \kappa\varepsilon)$ and $F_2(z^2 + \kappa\varepsilon)$

are defined and thus $r^o_{o,\varepsilon}$ is defined and analytic on $\mathcal{D}_\varepsilon$ by con-

struction.

In order to prove the bound in Lemma 6.2.4 we need some prep-

aratory estimates.

<u>Proposition 6.2.5.</u> <u>Let</u> $G_{o,\varepsilon} \in G_\varepsilon$. <u>The function</u> $G^2_{o,\varepsilon} = G_{o,\varepsilon} \circ G_{o,\varepsilon}$

<u>is defined and analytic on</u> $\mathcal{D}_\varepsilon$ <u>and we have the bound</u>

$$G^2_{o,\varepsilon}(z) = \lambda^2_\varepsilon \cdot z \cdot (1 - \frac{1+\lambda^2_\varepsilon}{\kappa}z^2 + \varepsilon z^3 B_\varepsilon(z) + z^4 B_\varepsilon(z)) \ .$$

<u>Proof</u> : For $G_{o,\varepsilon} \in G_\varepsilon$ we have $G_{o,\varepsilon}(\mathcal{D}_\varepsilon) \subset \mathcal{D}_\varepsilon$ by Proposition 6.2.2.

This proves that the function $G^2_{o,\varepsilon}$ (and in fact $G^n_{o,\varepsilon}$, $n > 0$;

n = n-fold iteration) is defined and analytic on $\mathcal{D}_\varepsilon$. The bound

follows then from the form of the Taylor expansion of $G^2_{o,\varepsilon}$ at

z = 0.

<u>Proposition 6.2.6.</u> <u>For</u> G_o , <u>with</u> $G_{o,\varepsilon} \in G_\varepsilon$, a_o , c_o , d_o , F_1

<u>and</u> F_2 <u>as defined in Definition 6.2.3, we have the equalities</u>

$$a_o((\frac{1}{G_o(z)})^2 - \frac{1}{z^2}) + c_o \log((\frac{G_o(z)}{z})^2) + d_o(G_o(z) - z)$$

$$+ F^\varepsilon_1((G_{o,\varepsilon}(z))^2) + G_{o,\varepsilon}(z)F^\varepsilon_2((G_{o,\varepsilon}(z))^2)$$

$$- F_1^\varepsilon(z^2) - zF_2^\varepsilon(z^2) - 1 = z \cdot \varepsilon \cdot B_\varepsilon(z) \qquad (6.2.6)$$

and

$$a_o((\frac{1}{G_o^2(z)})^2 - \frac{1}{z^2}) + c_o \log((\frac{G_o^2(z)}{z})^2) + d_o(G_o^2(z) - z)$$

$$+ F_1^\varepsilon((G_{o,\varepsilon}^2(z))^2) + G_{o,\varepsilon}^2(z)F_2^\varepsilon((G_{o,\varepsilon}^2(z))^2)$$

$$- F_1^\varepsilon(z^2) - zF_2^\varepsilon(z^2) - 2 = z^2 \varepsilon B_\varepsilon(z) + z\varepsilon^2 B_\varepsilon(z) \ .$$

$$(6.2.7)$$

Remark : If we take the formal limit $\varepsilon \to 0$ in (6.2.7) we get

Ecalle's equation and the equality gets trivially true. The crucial

point is that (6.2.7) expresses a relation between functions which

are defined not only on Ecalle's sectors, but on $\mathcal{D}_\varepsilon$, i.e. these

functions are analytic near $z = 0$.

Proof of Proposition 6.2.6 : To prove the proposition we split the

domain $\mathcal{D}_\varepsilon$ in d_ε and W_ε , and establish the necessary bounds

on both parts separately.

We first discuss equation (6.2.6) of Proposition 6.2.6 for

$z \in d_\varepsilon$.

We note, that by construction $-1 + a_o \cdot ((1/G_o(z))^2 - 1/z^2) +$

$+ c_o \cdot \log((G_o(z)/z)^2) + d_o \cdot (G_o(z) - z) = O(z^4)$, and that

$O(z^4) = z \cdot \varepsilon \mathcal{L}_\varepsilon(z)$ on d_ε. We show now that the terms containing

F_1 and F_2 satisfy such a bound, too. From the representation of

F_1 and F_2 by inverse Borel integrals (see Section 2.5) and using the definition (6.2.2) of S, it follows that the first order

derivatives F_1' and F_2' of F_1 and F_2 are analytic and bounded

functions on S. We consider now the decomposition

$$F_1^\epsilon(z^2) \equiv z^2 F_1(z^2 + \kappa \cdot \epsilon) = z^2 \cdot F_1(\kappa\epsilon) + z^2(F_1(z^2 + \kappa\epsilon) - F_1(\kappa\epsilon)).$$

The second summand can be locally bounded as follows

$$F_1(z^2 + \kappa\epsilon) - F_1(\kappa\epsilon) = \int_0^{z^2} F_1'(\kappa\epsilon + \xi)\,d\xi = \epsilon \ell_\epsilon(z)$$

where we have used the boundedness of F_1'. Using this bound we get

(with the abbreviation $H_o \equiv H_o(z)$)

$$z^2 \lambda_\epsilon^2 \cdot H_o^2 \cdot F_1(z^2 \lambda_\epsilon^2 H_o^2 + \kappa\epsilon) - z^2 F_1(z^2 + \kappa\epsilon)$$

$$= z^2(\lambda_\epsilon^2 - 1)H_o^2 F_1(z^2 \lambda_\epsilon^2 H_o^2 + \kappa\epsilon) + z^2 H_o^2 \cdot (F_1(z^2 \lambda_\epsilon^2 H_o^2 + \kappa\epsilon)$$

$$- F_1(\kappa\epsilon)) - z^2(F_1(z^2 + \kappa\epsilon) - F_1(\kappa\epsilon)) + F_1(\kappa\epsilon)z^2(H_o^2 - 1)$$

$$= \epsilon z^2 \ell_\epsilon(z) + z^4 \ell_\epsilon(z) = \epsilon z \ell_\epsilon(z) , \tag{6.2.8}$$

furthermore we have the bound

$$\epsilon F_2^\epsilon(z^2) = z^3 \cdot \ell_\epsilon(z) = \epsilon z \ell_\epsilon(z) \tag{6.2.9}$$

which leads to (in this case there are no cancellations present)

$$G_{o,\epsilon}(z)F_2^\epsilon((G_{o,\epsilon}(z))^2) - zF_2^\epsilon(z^2) = \epsilon z \ell_\epsilon(z). \tag{6.2.10}$$

We next estimate (6.2.6) on W_ε.

We have the decomposition

$$F_1^\varepsilon(z^2) \equiv z^2 F_1(z^2 + \kappa\varepsilon) = z^2 F_1(z^2) + z^2 \cdot (F_1(z^2 + \kappa\varepsilon) - F_1(z^2)) \ .$$

The second summand can be bounded as follows

$$F_1(z^2 + \kappa\varepsilon) - F_1(z^2) = \int_0^{\kappa\varepsilon} F_1'(z^2 + \xi)\,d\xi = \varepsilon W_\varepsilon(z) \ ,$$

and we get

$$F_1^\varepsilon(z^2) = z^2 F_1(z^2) + z^2 \cdot \varepsilon \cdot W_\varepsilon(z) \ . \tag{6.2.11}$$

A similar bound can be established for $F_2^\varepsilon(z^2)$.

Inserting these bounds in (6.2.6) we get the bound (using now the results of Section 2.5, i.e. the fact that we have solved Ecalle's equation)

$$a_o\left(\left(\frac{1}{G_o(z)}\right)^2 - \frac{1}{z^2}\right) + c_o \cdot \log\left(\left(\frac{G_o(z)}{z}\right)^2\right) + d_o(G_o(z) - z)$$

$$+ F_1^\varepsilon(\lambda_\varepsilon^2 z^2 H_o^2) + \lambda_\varepsilon z \cdot H_o F_2^\varepsilon(\lambda_\varepsilon^2 z^2 H_o^2)$$

$$- F_1^\varepsilon(z^2) - z F_2^\varepsilon(z^2) - 1 = z^2 \cdot \varepsilon W_\varepsilon(z) + z^3 \varepsilon W_\varepsilon(z) = z \varepsilon W_\varepsilon(z).$$

$$\tag{6.2.12}$$

Collecting (6.2.8) and (6.2.10) and glueing the bound so obtained

together with (6.2.12) provides the proof of (6.2.6) in Proposition 6.2.6.

To prove (6.2.7) we proceed in exactly the same way as for
(6.2.6) but use, that by Proposition 6.2.5 $G_{0,\varepsilon}^2(z) = \lambda_\varepsilon^2 \cdot z \cdot$
$\cdot (1 + z^2 \cdot B_\varepsilon(z))$. This leads to cancellations in addition to the
ones already explained. For $z \in \mathcal{D}_\varepsilon$ we can optimize the bound (6.2.10).
Namely, with $G_{0,\varepsilon}^2(z) \equiv z \cdot \lambda_\varepsilon^2 \cdot H_\varepsilon(z)$, we get

$$
G_{0,\varepsilon}^2(z) \cdot F_1^\varepsilon((G_{0,\varepsilon}^2(z))^2) - z F_2^\varepsilon(z^2) = (\lambda_\varepsilon^6 - 1) z^3 H_\varepsilon(z)^3 F_2(z^2 \lambda_\varepsilon^2 H_\varepsilon(z)^2 + \kappa\varepsilon)
$$

$$
+ z^3 H_\varepsilon(z)^3 (F_2(z^2 \lambda_\varepsilon^2 H_\varepsilon(z)^2 + \kappa\varepsilon) - F_2(\kappa\varepsilon))
$$

$$
- z^3 \cdot (F_2(z^2 + \kappa\varepsilon) - F_2(\kappa\varepsilon)) - F_2(\kappa\varepsilon) \cdot z^3 \cdot (H_\varepsilon(z)^3 - 1)
$$

$$
= \varepsilon z^3 \ell_\varepsilon(z) + z^5 \ell_\varepsilon(z) = \varepsilon z^2 \ell_\varepsilon(z) \ . \tag{6.2.13}
$$

Checking (6.2.8) and (6.2.12) we see that for those parts we have
already established the bounds which are necessary to prove (6.2.7).
Proposition 6.2.6 is proved.

We are now prepared to complete the proof of Lemma 6.2.4. In-
serting the definition of $F_{0,\varepsilon}^o$ and the definition of $G_{0,\varepsilon}(z) = $
$= \lambda_\varepsilon \cdot z \cdot H_o(z)$ into the r.h.s. of the equation which defines the
function $r_{0,\varepsilon}^o$, we get for $z \in \mathcal{D}_\varepsilon$

$$
r_{0,\varepsilon}^o(z) = \frac{z H_o}{\sqrt{z^2 \lambda^2 H_o^2 + \kappa\varepsilon}}(1 + \varepsilon c_o \log(1 + \frac{z^2 \lambda^2 H_o^2}{\kappa\varepsilon})) - \varepsilon c_o \frac{z^2 \lambda^2 H_o^2}{z^2 \lambda^2 H_o^2 + \kappa\varepsilon} +
$$

$$+ \varepsilon F_1^\varepsilon(z^2\lambda^2 H_o^2)) - \frac{z}{z^2+\kappa\varepsilon}\cdot(1 + \varepsilon c_o \log(1+\frac{z^2}{\kappa\varepsilon}) - \varepsilon c_o\frac{z^2}{z^2+\kappa\varepsilon} + \varepsilon F_1^\varepsilon(z^2)) +$$

$$+ \varepsilon\lambda^2\frac{z^3 H_o^3}{\sqrt{z^2\lambda^2 H_o^2+\kappa\varepsilon}^3}\cdot(d_o z\lambda H_o + \lambda z H_o F_2^\varepsilon(z^2\lambda^2 H_o^2)) - \frac{z^3}{\sqrt{z^2+\kappa\varepsilon}^3}\cdot$$

$$\cdot (d_o z + z F_2^\varepsilon(z^2\lambda^2 H_o^2)) .$$

We have used (and shall use) the abbreviation $\lambda \equiv \lambda_\varepsilon$. We introduce now the functions

$$Q(z) = \frac{z}{\sqrt{z^2+\kappa\varepsilon}}$$

and

$$(C(z))^2 = 1 - (Q(z))^2 = \frac{\kappa\varepsilon}{z^2+\kappa\varepsilon} = \kappa\varepsilon\frac{1}{z^2}(Q(z))^2 .$$

We shall use the abbreviations $Q \equiv Q(z)$, $C \equiv C(z)$. In addition
we define the function

$$X(z) = -1 + \frac{\kappa}{z^2}(\frac{1}{H_o^2} - 1) ,$$

and the function

$$Y(z) = Q(z)^2\cdot X(z) ,$$

and we shall use the abbreviation $X \equiv X(z)$ and $Y \equiv Y(z)$. In
terms of these functions we get for example

$$z^2\lambda^2 H_o^2 + \kappa\epsilon = H_o^2 \cdot (z^2 + \kappa\epsilon) \cdot (1 + \epsilon Y)$$

and

$$\log(1 + \frac{z^2\lambda^2 H_o^2}{\kappa\epsilon}) = \log H_o^2 + \log(1 + \epsilon Y) + \log(1 + \frac{z^2}{\kappa\epsilon}) \, ,$$

and for the function $r_{o,\epsilon}^o$ we calculate

$$r_{o,\epsilon}^o(z) = Q \cdot [(1 + \epsilon Y)^{-1/2} - 1] \cdot [1 + \epsilon c_o \log(H_o^2) + \epsilon c_o \log(1 + \epsilon Y)$$

$$+ \epsilon c_o \log(1 + \frac{z^2}{\kappa\epsilon}) - \epsilon c_o \lambda^2 Q^2 (1 + \epsilon Y)^{-1} + \epsilon F_1^\epsilon (z^2\lambda^2 H_o^2)]$$

$$- \epsilon Q^3 c_o \cdot (\lambda^2 - 1) + \epsilon Q \cdot [c_o \log H_o^2 + c_o \log(1 + \epsilon Y)$$

$$- c_o (\lambda^2 - 1) Q^2 \cdot [(1 + \epsilon Y)^{-1} - 1] - c_o \Omega^2 \cdot [(1 + \epsilon Y)^{-1} - 1]$$

$$+ F_1^\epsilon (z^2\lambda^2 H_o^2) - F_1^\epsilon (z^2)] + \epsilon Q^3 \cdot (\lambda^2 - 1) \cdot (1 + \epsilon Y)^{-3/2}$$

$$\cdot (d_o \lambda z H_o + \lambda z H_o F_2^\epsilon (z^2\lambda^2 H_o^2)) + \epsilon Q^3 \cdot [(1 + \epsilon Y)^{-3/2} - 1]$$

$$\cdot (d_o z \lambda H_o + \lambda z H_o F_2^\epsilon (z^2\lambda^2 H_o^2)) + \epsilon Q^3 \cdot [d_o \cdot z \cdot (\lambda H_o - 1)$$

$$+ \lambda z H_o F_2^\epsilon (z^2\lambda^2 H_o^2) - z F_2^\epsilon (z^2)] \equiv \sum_{i=1}^{\infty} R_i(z) \, ,$$

and we have used the definitions

$$R_1(z) = Q \cdot [(1 + \epsilon Y)^{-1/2} - 1 + \frac{1}{2}\epsilon Y] \, ,$$

$$R_2(z) = \varepsilon Q \cdot [(1 + \varepsilon Y)^{-1/2} - 1] \cdot [c_o \cdot \log(H_o^2) + c_o \cdot \log(1 + \varepsilon Y)$$

$$- c_o \lambda^2 \Omega^2 (1 + \varepsilon Y)^{-1} + F_1^\varepsilon(z^2 \lambda^2 H_o^2)] \ ,$$

$$R_3(z) = \varepsilon Q \cdot [(1 + \varepsilon Y)^{-1/2} - 1] \cdot c_o \cdot \log(1 + z^2/\kappa\varepsilon) \ ,$$

$$R_4(z) = \varepsilon Q^3 \cdot [-\tfrac{1}{2}X + c_o \log(H_o^2) + F_1^\varepsilon(z^2 \lambda^2 H_o^2) - F_1^\varepsilon(z^2) + d_o z \lambda H_o$$

$$- d_o z + \lambda z H_o F_2^\varepsilon(z^2 \lambda^2 H_o^2) - z F_2^\varepsilon(z^2)] \ ,$$

$$R_5(z) = \varepsilon Q \cdot [c_o \log(1 + \varepsilon Y) - c_o(\lambda^2 - 1) Q^2 \cdot [(1 + \varepsilon Y)^{-1} - 1]$$

$$- c_o \Omega^2 \cdot [(1 + \varepsilon Y)^{-1} - 1]] \ ,$$

$$R_6(z) = \varepsilon Q \cdot c^2 \cdot c_o \log(H_o^2) - \varepsilon Q^3 c_o(\lambda^2 - 1) \ ,$$

$$R_7(z) = \varepsilon Q \cdot c^2 [F_1^\varepsilon(z^2 \lambda^2 H_o^2) - F_1^\varepsilon(z^2)] \ ,$$

$$R_8(z) = \varepsilon Q^3 [(1 + \varepsilon Y)^{-3/2} - 1] \cdot (d_o \cdot z \cdot \lambda \cdot H_o + \lambda z H_o F_2^\varepsilon(z^2 \lambda^2 H_o^2) \ .$$

The idea is now the following. With the exception of R_4, the functions R_i trivially satisfy the bound required in Lemma 6.2.4. The real problem is therefore to show that the function R_4 satisfies such a bound, too. But the function R_4 is essentially zero because a_o, c_o, d_o and F_1 and F_2 solve Ecalle's equation by the construction in Section 5. All the dangerous terms cancel by construction. This justifies from an other point of view our heuristic arguments given in the introduction to this book.

We establish now the bound we need to prove Lemma 6.2.4 for each of these eight terms separately. First note that

$$X(z) = -2 + \frac{\kappa}{z^2}((\frac{1}{H_o(z)})^2 - 1) = -4d_o \cdot z - \frac{4}{\kappa} \cdot c_o \cdot z^2 + z^3 B_\varepsilon(z)$$

$$(6.2.14)$$

by definition of κ, c_o, and d_o, and that therefore

$$Y(z) \equiv Q(z)^2 \cdot X(z) \equiv zB_\varepsilon(z) .$$

Using this we get for the function R_1,

$$R_1(z) = Q \cdot ((1 + \varepsilon Y)^{-1/2} - 1 + \frac{1}{2}\varepsilon Y) = Q\frac{1}{\sqrt{1+\varepsilon Y}} \cdot (1 - (1 + \varepsilon Y)^{1/2}$$

$$\cdot (1 - \frac{1}{2}\varepsilon Y)) = Q \cdot \frac{1}{\sqrt{1+\varepsilon Y}} \cdot (1 + (1 + \varepsilon Y)^{1/2} \cdot (1 - \frac{1}{2}\varepsilon Y))^{-1}$$

$$\cdot (1 - \frac{1}{2}\varepsilon Y)^2) = Q\frac{1}{\sqrt{1+\varepsilon Y}} \cdot (1 + \sqrt{1+\varepsilon Y}(1 - \frac{1}{2}\varepsilon Y))^{-1}$$

$$\cdot (-(1 + \varepsilon Y)\frac{1}{4}\varepsilon^2 Y^2 + \varepsilon^2 Y^2) = \varepsilon^2 Q^5 x^2 (1 + \varepsilon Y)^{-1/2} \cdot (1 - (1+\varepsilon Y)$$

$$+ (1 + \varepsilon Y)^{1/2}(1 - \frac{1}{2}\varepsilon Y))^{-1}(1 - \frac{1}{4}(1 + \varepsilon Y)) = \varepsilon^2 Q^5 z^2 B_\varepsilon(z)$$

$$= \varepsilon^2 Q^3 z^2 B_\varepsilon(z) = \varepsilon^2 \cdot \log\varepsilon Q^3 B_\varepsilon(z)$$

as required. For the function R_2 we get similarly, inspecting term by term

$$R_2(z) = \varepsilon^2 Q^3 zB_\varepsilon(z)[z^2 B_\varepsilon(z) + \varepsilon Q^2 zB_\varepsilon(z) + Q^2 B_\varepsilon(z) + z^2 B_\varepsilon(z)]$$

$$= \varepsilon Q^3 z B_\varepsilon (z) = \varepsilon^2 Q^3 B_\varepsilon (z) = \varepsilon^2 \cdot \log \varepsilon \cdot Q^3 B_\varepsilon (z)$$

as required, and for the function R_3

$$R_3 (z) = \varepsilon^2 Q^3 z B_\varepsilon (z) \log (1 + z^2/\kappa \varepsilon) = \varepsilon^2 \cdot \log \varepsilon \cdot Q^3 \cdot z \cdot B_\varepsilon (z) =$$

$$= \varepsilon^2 \log \varepsilon Q^3 B_\varepsilon (z)$$

as required. In function R_4 we replace X by the expansion
(6.2.14) and use then (6.2.6) of Proposition 6.2.6 to establish

$$R_4 (z) = \varepsilon^2 Q^3 z B_\varepsilon (z) = \varepsilon^2 \log \varepsilon \cdot Q^3 B_\varepsilon (z)$$

as required. For the functions R_5, R_6, R_7 and R_8 we get successively, inspecting term by term

$$R_5 (z) = \varepsilon^2 Q^3 z [B_\varepsilon (z) + \varepsilon Q^2 B_\varepsilon (z) + Q^2 B_\varepsilon (z)] = \varepsilon^2 Q^3 z B_\varepsilon (z) =$$

$$= \varepsilon^2 \cdot \log \varepsilon Q^3 B_\varepsilon (z) \ ,$$

$$R_6 (z) = \varepsilon^2 Q^3 \kappa c_o \frac{1}{z^2} (-\frac{2}{\kappa} z^2 + z^3 B_\varepsilon (z)) + 2\varepsilon^2 c_o Q^3$$

$$= \varepsilon^2 Q^3 z B_\varepsilon (z) = \varepsilon^2 \log \varepsilon \cdot Q^3 B_\varepsilon (z) \ ,$$

$$R_7 (z) = \varepsilon^2 Q^3 \kappa \frac{1}{z^2} [z^2 \lambda^2 H_o^2 \cdot F_1 (z^2 \lambda^2 H_o^2 + \kappa \varepsilon) - z^2 F_1 (z^2 + \kappa \varepsilon)]$$

$$+ \varepsilon^2 Q^3 \cdot \kappa \cdot [(\lambda^2 - 1) H_o^2 F_1 (z^2 \lambda^2 H_o^2 + \kappa \varepsilon) + H_o^2 F_1 (z^2 \lambda^2 H_o^2 + \kappa \varepsilon)$$

$$- F_1 (z^2 + \kappa \varepsilon)] = \varepsilon^2 Q^3 (\varepsilon B_\varepsilon(z) + z^2 B_\varepsilon(z)) = \varepsilon^2 Q^3 B_\varepsilon(z)$$

$$= \varepsilon^2 \log \varepsilon Q^3 B_\varepsilon(z) \ ,$$

$$R_8(z) = \varepsilon^2 Q^3 \cdot z \cdot (z B_\varepsilon(z) + z^3 B_\varepsilon(z)) + \varepsilon^2 z^2 Q^5 B_\varepsilon(z)$$

$$= \varepsilon^2 \log \varepsilon \cdot Q^3 B_\varepsilon(z)$$

as required. This completes the proof of Lemma 6.2.4.

The next lemma shows, that there is a function $F_{0,\varepsilon}$ which solves the Schröder equation on the domain $\mathcal{D}_\varepsilon$.

Lemma 6.2.7. Let $G_{0,\varepsilon} \in G_\varepsilon$ and let $r^0_{0,\varepsilon}$ be the function defined in Lemma 6.2.4. For ε sufficiently small, but fixed, the sum

$$F^>_{0,\varepsilon}(z) = - \sum_{n=0}^{\infty} (\frac{1}{\lambda_\varepsilon})^n \cdot r^0_{0,\varepsilon} (G^n_{0,\varepsilon}(z))$$

converges uniformly on $\mathcal{D}_\varepsilon$, defines therefore a function which is analytic on $\mathcal{D}_\varepsilon$, and the function

$$F_{0,\varepsilon}(z) = F^0_{0,\varepsilon}(z) + F^>_{0,\varepsilon}(z)$$

is the (up to a trivial multiplicative factor) unique function de-fined and analytic on $\mathcal{D}_\varepsilon$ which solves (6.1.11).

We have to estimate how fast the iterates $G^n_{0,\varepsilon}(z_0)$ approach $z = 0$, for $z_0 \in \mathcal{D}_\varepsilon$. From the bound

$$\mathrm{Re}\left(\left(\frac{1}{G_o(z)}\right)^2\right) \geq \mathrm{Re}\left(\frac{1}{z^2}\right) + L , \qquad L > 0 ,$$

which we have already established in (5.7.6), we get

$$\mathrm{Re}\left(\left(\frac{1}{G_{o,\varepsilon}(z)}\right)^2\right) \geq \left(\frac{1}{\lambda_\varepsilon}\right)^2 \left(\mathrm{Re}\left(\frac{1}{z^2}\right) + L\right) .$$

Iterating this inequality once we get

$$\mathrm{Re}\left(\left(\frac{1}{G_{o,\varepsilon}^2(z)}\right)^2\right) \geq \left(\frac{1}{\lambda_\varepsilon}\right)^4 \left(\mathrm{Re}\left(\frac{1}{z^2}\right) + (1 + (\lambda_\varepsilon)^2)L\right)$$

and after n-fold iteration

$$\mathrm{Re}\left(\left(\frac{1}{G_{o,\varepsilon}^n(z)}\right)^2\right) \geq \left(\frac{1}{(\lambda_\varepsilon)^2}\right)^n \cdot \left(\mathrm{Re}\left(\frac{1}{z^2}\right) + L \sum_{m=0}^{n-1} ((\lambda_\varepsilon)^2)^m\right)$$

$$= \left(\frac{1}{(\lambda_\varepsilon)^2}\right)^n \cdot \left(\mathrm{Re}\left(\frac{1}{z^2}\right) + L \cdot \left(\frac{1-((\lambda_\varepsilon)^2)^n}{1 - (\lambda_\varepsilon)^2}\right)\right)$$

$$= \left(\frac{1}{(\lambda_\varepsilon)^2}\right)^n \cdot \left(\frac{L_6}{|z|^2} + L\left(\frac{1-((\lambda_\varepsilon)^2)^n}{1 - (\lambda_\varepsilon)^2}\right)\right) ,$$

where $L_6 = \cos 2\varphi_o$. (φ_o is the opening angle of the sectors $S^{\pm}$.)
Finally we get

$$|G_{o,\varepsilon}^n(z)| \leq |\lambda_\varepsilon|^n \left(L \left(\frac{1-(\lambda_\varepsilon)^{2n}}{1-(\lambda_\varepsilon)^2}\right) + \frac{L_6}{|z|^2}\right)^{-1/2} . \qquad (6.2.15)$$

We can now estimate (using the bound which we have established in

Lemma 6.2.4)

$$\left| r^o_{o,\varepsilon} (G^n_{o,\varepsilon} (z)) \right| \leq \text{const.} \varepsilon^2 |\log\varepsilon| \cdot |\lambda_\varepsilon|^{3n}$$

$$\frac{(L \cdot (\dfrac{1-(\lambda_\varepsilon)^{2n}}{1 - (\lambda_\varepsilon)^2}) + L_6 \dfrac{1}{|z|^2})^{-3/2}}{[(\lambda_\varepsilon)^{2n}(L \cdot (\dfrac{1-(\lambda_\varepsilon)^{2n}}{1 - (\lambda_\varepsilon)^2}) + L_6 \dfrac{1}{|z|^2})^{-1} + \kappa\varepsilon]^{3/2}}$$

$$\leq \text{const.} \varepsilon^2 |\log\varepsilon| \cdot |\lambda_\varepsilon|^{3n},$$

and therefore for $z \in \mathcal{D}_\varepsilon$

$$\left| F^>_{o,\varepsilon} (z) \right| \leq \text{const.} \sum_{n=0}^{\infty} \varepsilon^2 |\log\varepsilon| \cdot |\lambda_\varepsilon|^{2n}$$

$$= \text{const.} \varepsilon^2 |\log\varepsilon| \cdot \frac{1}{1-(\lambda_\varepsilon)^2} \leq \text{const.} \varepsilon |\log\varepsilon|$$

which proves that the sum defining $F^>_{o,\varepsilon}$ is uniformly convergent. That the function $F_{o,\varepsilon}$ solves the Schröder equation is clear by construction, and it remains the uniqueness to be proved. Suppose, that we are given two functions f_1 and f_2 both not identically zero, and both verifying

$$(\frac{1}{\lambda_\varepsilon}) \cdot f_\delta (G_{o,\varepsilon} (z)) - f_\delta (z) , \qquad \delta = 1,2 .$$

We have $f_1 (0) = f_2 (0) = 0$ and we may suppose (without loss of generality) $f_1' (0) = f_2' (0) = 1$. By linearity the function $f = f_1 - f_2$ is again a solution and $f(z) = 0(z^2)$. But then

$$f(z) = \frac{1}{\lambda_\varepsilon} f (G_{o,\varepsilon} (z)) = (\frac{1}{\lambda_\varepsilon})^n \cdot f (G^n_{o,\varepsilon} (z)) = 0((\lambda_\varepsilon)^n)$$

which approaches zero, when $n \to \infty$ for any $z \in \mathcal{D}_\varepsilon$, a contradiction. Lemma 6.2.7 is proved.

Lemma 6.2.8. Let $F^>_{o,\varepsilon}$ be the function given by Lemma 6.2.7. It obeys the bound

$$F^>_{o,\varepsilon}(z) = \varepsilon^2 (\log \varepsilon)^2 \cdot B_\varepsilon(z) .$$

Proof : In Lemma 6.2.7 we have already proved the bound

$$F^>_{o,\varepsilon}(z) = \varepsilon \cdot \log \varepsilon \cdot B_\varepsilon(z)$$

which is not good enough for our needs, and we have to sharpen the estimates. We proceed as follows. Because the sum defining the function $F^>_{o,\varepsilon}$, namely

$$F^>_{o,\varepsilon}(z) = - \sum_{n=0}^{\infty} (\frac{1}{\lambda_\varepsilon})^n r^o_{o,\varepsilon} (G^n_{o,\varepsilon}(z))$$

is uniformly convergent, we may replace it equivalently by the sum

$$F^>_{o,\varepsilon}(z) = - \sum_{n=0}^{\infty} (\frac{1}{\lambda_\varepsilon})^{2n} (r^o_{o,\varepsilon} (G^{2n}_{o,\varepsilon}(z)) + \frac{1}{\lambda_\varepsilon} r^o_{o,\varepsilon} (G^{2n+1}_{o,\varepsilon}(z)))$$

for which we can establish better bounds, using to our favor the explicit cancellations which occur. We define first the function $r_{o,\varepsilon}$ by the equation

$$r_{o,\varepsilon}(z) = r^o_{o,\varepsilon}(z) + \frac{1}{\lambda_\varepsilon} r^o_{o,\varepsilon} (G_{o,\varepsilon}(z)) .$$

Using now the definition of the function $r^o_{o,\varepsilon}$ we get

$$r_{o,\varepsilon}(z) = (\frac{1}{\lambda_\varepsilon})^2 F^o_{o,\varepsilon}(G^2_{o,\varepsilon}(z)) - F^o_{o,\varepsilon}(z) .$$

We use now the bound of Proposition 6.2.5 and proceed exactly as in the proof of Lemma 6.2.4. We use the notation $G^2_{o,\varepsilon}(z) = \lambda^2_\varepsilon \cdot z \cdot H_\varepsilon(z)$ (which defines $H_\varepsilon = H_\varepsilon(z)$), $X_\varepsilon = X_\varepsilon(z) = -4 + + 4\varepsilon + \frac{\kappa}{z^2}((1/H_\varepsilon)^2 - 1)$ and $Y_\varepsilon = Y_\varepsilon(z) = Q^2(z) \cdot X_\varepsilon$. The function X_ε obeys the bound

$$X_\varepsilon(z) = -4 + 4\varepsilon + \frac{\kappa}{z^2}(1 + (\frac{2-\varepsilon}{\kappa})z^2 + \varepsilon z^3 B_\varepsilon(z) + z^4 B_\varepsilon(z) - 1)$$

$$= \varepsilon \cdot z \cdot B_\varepsilon(z) + z^2 B_\varepsilon(z) .$$

We get the decomposition

$$r_{o,\varepsilon}(z) = \sum_{i=1}^{8} R^\varepsilon_i(z)$$

where the functions R_i are defined by the equations

$$R^\varepsilon_1(z) = Q \cdot [(1 + \varepsilon Y_\varepsilon)^{-1/2} - 1 + \frac{1}{2}\varepsilon Y_\varepsilon] ,$$

$$R^\varepsilon_2(z) = \varepsilon Q \cdot [(1 + \varepsilon Y_\varepsilon)^{-1/2} - 1] \cdot [c_o \cdot \log H^2_\varepsilon + c_o \log(1 + \varepsilon Y_\varepsilon)$$

$$- c_o \lambda^4 Q^2 (1 + \varepsilon Y_\varepsilon)^{-1} + F^\varepsilon_1(z^2 \lambda^4 H^2_\varepsilon)] ,$$

$$R^\varepsilon_3(z) = \varepsilon Q \cdot [(1 + \varepsilon Y_\varepsilon)^{-1/2} - 1]c_o \cdot \log(1 + z^2/\kappa\varepsilon) ,$$

$$R_4^\epsilon(z) = \epsilon Q^3 \cdot [-\tfrac{1}{2}X_\epsilon + c_o \cdot \log H_\epsilon^2 + F_1^\epsilon(z^2\lambda^2 H_\epsilon^2) - F_1^\epsilon(z^2) + d_o z\lambda^2 H_\epsilon$$

$$- d_o z + \lambda^2 z H_\epsilon F_2^\epsilon(z^2\lambda^4 H_\epsilon^2) - z F_2^\epsilon(z^2)] \ ,$$

$$R_5^\epsilon(z) = \epsilon Q \cdot [c_o \cdot \log(1 + \epsilon Y_\epsilon) - c_o(\lambda^4 - 1)Q^2[(1 + \epsilon Y_\epsilon)^{-1} - 1]$$

$$- c_o Q^2[(1 + \epsilon Y_\epsilon)^{-1} - 1]] \ ,$$

$$R_6^\epsilon(z) = \epsilon Q C^2 c_o \cdot \log(H_\epsilon^2) - \epsilon Q^3 c_o(\lambda^4 - 1) \ ,$$

$$R_7^\epsilon(z) = \epsilon Q C^2 \cdot [F_1^\epsilon(z^2\lambda^4 H_\epsilon^2) - F_1^\epsilon(z^2)] \ ,$$

$$R_8^\epsilon(z) = \epsilon Q^3[(1 + \epsilon Y_\epsilon)^{-3/2} - 1](d_o\lambda^2 z H_\epsilon + \lambda^2 z H_\epsilon F_2^\epsilon(z^2\lambda^4 H_\epsilon^2)) \ .$$

We continue again as in the proof of Lemma 6.2.4 and give bounds
for each of these eight terms separately but we only indicate how
the terms have to be estimated and do not discuss everything in
full detail. We have

$$R_1^\epsilon(z) = Q\epsilon^2 Y_\epsilon^2 B_\epsilon(z) = \epsilon^2 Q^5 X_\epsilon^2 B_\epsilon(z) = \epsilon^2 Q^3(\epsilon \cdot z B_\epsilon(z) + z^2 B_\epsilon(z))^2$$

$$= \epsilon^2 z^2 Q^3 B_\epsilon(z) \ ,$$

$$R_2^\epsilon(z) = \epsilon^2 Q^3(\epsilon z B_\epsilon(z) + z^2 B_\epsilon(z))(z^2 B_\epsilon(z) + \epsilon^2 z B_\epsilon(z)$$

$$+ \epsilon z^2 B_\epsilon(z) + B(z) + z^2 B(z)) = \epsilon Q^3 z^2 B_\epsilon(z) + \epsilon Q^3 \cdot z \cdot B_\epsilon(z)$$

$$= \epsilon^2 Q^3(z^2 B_\epsilon(z) + \epsilon z B_\epsilon(z)) \ ,$$

$$R_3^\varepsilon(z) = \varepsilon^2 Q^3 \cdot (\varepsilon z B_\varepsilon(z) + z^2 B_\varepsilon(z)) \cdot \log(1 + z^2/\kappa\varepsilon)$$

$$= \varepsilon^2 \log\varepsilon \cdot Q^3 (z^2 B_\varepsilon(z) + \varepsilon z B_\varepsilon(z)) \ .$$

For the function R_4^ε we use (6.2.7) of Proposition 6.2.6 to conclude that

$$R_4^\varepsilon(z) = \varepsilon Q^3 (z^2 \cdot \varepsilon B_\varepsilon(z) + z \cdot \varepsilon^2 B_\varepsilon(z)) = z^2 \cdot \varepsilon^2 Q^3 (z^2 \cdot B_\varepsilon(z)$$

$$+ \ \varepsilon z B_\varepsilon(z)) \ .$$

Finally we have

$$R_5^\varepsilon(z) = \varepsilon Q \cdot [Q^2 \cdot \varepsilon (\varepsilon z B_\varepsilon(z) + z^2 B_\varepsilon(z))] = \varepsilon^2 Q^3 \cdot (z^2 B_\varepsilon(z)$$

$$+ \ \varepsilon z B_\varepsilon(z)) \ ,$$

$$R_6^\varepsilon(z) = \varepsilon^2 Q^3 \kappa c_0 \frac{1}{z^2} \cdot \log(H_\varepsilon^2) - \varepsilon Q^3 c_0 (\lambda^2 - 1)$$

$$= \varepsilon^2 Q^3 \cdot c_0 (\frac{\kappa}{z^2} \cdot \log(1 - 2(\frac{2-\varepsilon}{\kappa}) z^2 + \varepsilon z^3 B_\varepsilon(z) + z^4 B_\varepsilon(z))$$

$$- \frac{1}{\varepsilon}((1 - 2\varepsilon)^2 - 1)) = \varepsilon^2 Q^3 c_0 (-4 + 4\varepsilon + \varepsilon z B_\varepsilon(z)$$

$$+ \ z^2 B_\varepsilon(z) + 4 - 4\varepsilon) = \varepsilon^2 Q^3 \cdot (z^2 B_\varepsilon(z) + \varepsilon z B_\varepsilon(z)) \ ,$$

$$R_7^\varepsilon(z) = \varepsilon^2 Q^3 z^2 B_\varepsilon(z)$$

and

$$R_8^\epsilon(z) = \epsilon Q^3 \cdot [\epsilon(\epsilon \cdot zB_\epsilon(z) + z^2 B_\epsilon(z))] \cdot z \cdot B_\epsilon(z) = \epsilon^2 Q^3 z^2 B_\epsilon(z) \ .$$

Collecting these bounds together we find that we have proved the following proposition.

Proposition 6.2.9. The function $r_{o,\epsilon}^o$ obeys the bound

$$r_{o,\epsilon}^o(z) = \epsilon^2 \cdot \log\epsilon \cdot Q^3 \cdot (z^2 B_\epsilon(z) + \epsilon z B_\epsilon(z)) \ .$$

We can now complete the proof of Lemma 6.2.8. We split once more the sum defining $F_{o,\epsilon}^>$, namely we write

$$F_{o,\epsilon}^>(z) = - \sum_{n=0}^\infty (\frac{1}{\lambda_\epsilon})^{2n} \cdot r_{o,\epsilon} (G_{o,\epsilon}^{2n}(z)) \equiv \Sigma_1(z) + \Sigma_2(z) \ ,$$

where the functions Σ_1 and Σ_2 are defined such a way that, using Proposition 6.2.9, we get

$$\Sigma_1(z) = \epsilon^2 \cdot \log\epsilon \cdot \sum_{n=0}^\infty (\frac{1}{\lambda_\epsilon})^{2n} \cdot [Q^3 z^2 B_\epsilon(z)] \circ G_{o,\epsilon}^{2n}(z) \ ,$$

$$\Sigma_2(z) = \epsilon^3 \cdot \log\epsilon \cdot \sum_{n=0}^\infty (\frac{1}{\lambda_\epsilon})^{2n} \cdot [Q^3 \cdot z \cdot B_\epsilon(z)] \circ G_{o,\epsilon}^{2n}(z) \ .$$

Using now (6.2.15), we see, that for $z \epsilon \mathcal{D}_\epsilon$

$$|[Q^3 z^2 B_\epsilon(z)] \circ G_{o,\epsilon}^{2n}(z)| \leq const. |\lambda_\epsilon|^{10n} \cdot (L(\frac{1-(\lambda_\epsilon)^{4n}}{1-(\lambda_\epsilon)^4}) + L_6 \frac{1}{|z|^2})^{-1}$$

$$\leq const. |\lambda_\epsilon|^{10n} \frac{|z|^2 \cdot \epsilon}{|z|^2 (1-(\lambda_\epsilon)^{4n}) + L \cdot \epsilon} \leq const. |\lambda_\epsilon|^{10n} \cdot \frac{\epsilon}{1-(\lambda_\epsilon)^{4n}}$$

and

$$| [Q^3 z_\varepsilon (z)] \circ G^{2n}_{o,\varepsilon} (z) | \leq \text{const.} |\lambda_\varepsilon|^{8n} (L \cdot (\frac{1 - (\lambda_\varepsilon)^{4n}}{1 - (\lambda_\varepsilon)^4}) + L_6 \frac{1}{|z|^2})^{-1/2}$$

$$\leq \text{const.} |\lambda_\varepsilon|^{8n} \frac{\varepsilon^{1/2}}{(|z|^2 (1-(\lambda_\varepsilon)^{4n}) + \tilde{L} \cdot \varepsilon)^{1/2}} \leq |\lambda_\varepsilon|^{8n} \frac{\varepsilon^{1/2}}{(1-(\lambda_\varepsilon)^{4n})^{1/2}} .$$

We can now estimate the function Σ_1 (C_1, C_2,... are constants independent of ε)

$$\sup_{z \in \mathcal{D}_\varepsilon} |\Sigma_1 (z)| \leq C_1 \cdot \varepsilon^2 |\log\varepsilon| + \varepsilon^3 \cdot |\log\varepsilon| \cdot C_2 \cdot \sum_{n=2}^{\infty} \frac{|\lambda_\varepsilon|^{8n}}{1-(\lambda_\varepsilon)^{4n}}$$

$$\leq C_1 \cdot \varepsilon^2 \cdot |\log\varepsilon| + \varepsilon^3 \cdot |\log\varepsilon| \cdot C_2 \cdot \int_4^{\infty} \frac{|\lambda_\varepsilon|^{2x}}{1-|\lambda_\varepsilon|^x} dx$$

$$\leq C_3 \cdot \varepsilon^2 (\log\varepsilon)^2$$

and similarly for the function Σ_2

$$\sup_{z \in \mathcal{D}_\varepsilon} |\Sigma_2 (z)| \leq C_4 \cdot \varepsilon^{7/2} |\log\varepsilon| + \varepsilon^{7/2} |\log\varepsilon| \cdot C_3 \cdot \int_4^{\infty} \frac{|\lambda_\varepsilon|^{3/2 \cdot x}}{(1-|\lambda_\varepsilon|^x)^{1/2}} dx$$

$$\leq C_6 \cdot \varepsilon^{5/2} |\log\varepsilon|$$

Lemma 6.2.8 is proved.

Remark : The bound for the function $F^>_{o,\varepsilon}$ given in Lemma 6.2.8 is still not optimal. It can be sharpened to

$$F^>_{o,\varepsilon} = \varepsilon^2 \cdot \log\varepsilon \cdot B_\varepsilon (z) ,$$

but because the estimates needed to prove Lemma 6.2.8 are already quite cumbersome and since we do not need this (optimal) bound, we do not insist on this point.

We can now complete the proof of Theorem 6.2.1. First note, that the transformation $T_{-A}(z) = z/(1 - Az)$ maps the disc $D(-x_{oh}, \rho_h) = \{z \in \mathbb{C} \mid |z + x_{oh}| < \rho_h\}$ onto some other disc. A little lemma (which we have also proved with our computer program, see estimates following Fig. 5.7.2) shows that

$$T_{-A}(D(-x_{oh}, \rho_h)) \subset \mathcal{D}_\varepsilon , \qquad (6.2.16)$$

for some discs $D(-x_{oh}, \hat{\rho}_h)) = \{z \in \mathbb{C} \mid |z + x_{oh}| < \hat{\rho}_h ; \hat{\rho}_h > \rho_h\}$. Using now the explicit form of $F_{o,\varepsilon}^o$ and Lemma 6.2.8 we see that $F_{o,\varepsilon}(T_{-A}(-x_{oh})) = 1 + \text{const.} \cdot \varepsilon \cdot \log\varepsilon$, which is >0 and therefore the function h_ε^u as defined through (6.1.13) is analytic on the unit disc. It remains the bound (6.2.1) to be proved. We fix some $\hat{\rho} > \rho$ for which (6.2.16) is satisfied. Using again the approxima-tion $F_{o,\varepsilon}^o$ of $F_{o,\varepsilon}$ constructed in this section we get that for $z \in T_{-A}(D(-x_{oh}, \hat{\rho}_h))$

$$F_{o,\varepsilon}^o(z) = (1 - \tfrac{1}{2}\kappa\varepsilon/z^2 + \varepsilon^2 \cdot B_h(z)) \cdot F_{1,\varepsilon}(z^2) + \varepsilon z \cdot$$

$$\cdot (1 + \varepsilon B_h(z)) F_{2,\varepsilon}(z^2) ,$$

where B_h denotes functions, bounded and analytic on $T_{-A}(D(-x_{oh}, \hat{\rho}_h))$. Next we note that, using Definition 6.2.3,

$$F_{1,\varepsilon}(z^2) = 1 - \varepsilon c_o \cdot \log(\kappa\varepsilon) - \varepsilon c_o + \varepsilon c_o \cdot \log(z^2) + \varepsilon F_1^\varepsilon(z^2)$$

$$+ \varepsilon^2 B_h(z) ,$$

and therefore

$$F_{o,\varepsilon}^o(z) = 1 - \varepsilon c_o \log(\kappa\varepsilon) - \varepsilon \cdot c_o + a_o \varepsilon/z^2 + \varepsilon \cdot c_o \cdot \log(z^2)$$

$$+ \varepsilon F_1^\varepsilon(z^2) + \varepsilon^2 \cdot \log\varepsilon B_h(z) + \varepsilon z(d_o + F_2^\varepsilon(z^2)) + \varepsilon^2 z B_h(z)$$

$$= (1 - \varepsilon c_o \log(\kappa\varepsilon) - \varepsilon c_o) + \varepsilon[a_o/z^2 + c_o \cdot \log(z^2)$$

$$+ z^2 F_1(z^2) + d_o z + z^3 F_2(z^2)] = \varepsilon^2 \cdot \log\varepsilon \cdot B_h(z) .$$

We can identify the function in the bracket with F_o constructed in Section 5 of this paper. Using in addition the bound for the function $F_{o,\varepsilon}^>$ which we have established in Lemma 6.2.8 we get

$$F_{o,\varepsilon}(z) = F_{o,\varepsilon}^o(z) + F_{o,\varepsilon}^>(z) = 1 - \varepsilon \cdot c_o \cdot \log(\kappa\varepsilon) - \varepsilon c_o$$

$$+ \varepsilon F_o(z) + \varepsilon^2 \cdot (\log\varepsilon)^2 B_h(z) ,$$

and therefore, for z in the unit disc

$$h_\varepsilon^u(\frac{\hat{\rho}_h}{\rho_h} z) = \frac{1-\varepsilon \cdot c_o \cdot \log(\kappa\varepsilon) - \varepsilon c_o + \varepsilon F_o(T_{-A}(-x_{oh} + \hat{\rho}_h z)) + \varepsilon^2(\log\varepsilon)^2 B_u(z)}{1-\varepsilon \cdot c_o \cdot \log(\kappa\varepsilon) - \varepsilon c_o + \varepsilon F_o(T_{-A}(-x_{oh})) + const. \varepsilon^2 (\log\varepsilon)^2} - 1$$

$$= \varepsilon \cdot [F_o(T_{-A}(-x_{oh} + \hat{\rho}_h z)) - F_o(T_{-A}(-x_{oh}))] + \varepsilon^2(\log(\varepsilon))^2 B_u(z)$$

$$= h^u(\frac{\hat{\rho}_h}{\rho_h}) + \varepsilon^2(\log\varepsilon)^2 B_u(z) ,$$

where B_u denotes functions, bounded and analytic on the unit disc.
Using now Cauchy's formula to estimate the Taylor coefficients at
$z = 0$ leads to

$$\| h_\varepsilon^u(.) - \varepsilon h^u(.) \| \leq \frac{1}{1-(\rho_h/\tilde\rho_h)} \cdot \varepsilon^2 \cdot (\log\varepsilon)^2 \cdot \sup_{\substack{z \in \\ \text{unit disc}}} |B_u(z)|$$

$$\leq \text{const.} \varepsilon^2 \cdot (\log\varepsilon)^2 .$$

Theorem 6.2.1 is proved.

This theorem is the key step in the proof of Theorem 6.1.2.
Given the above bound for h_ε^u one constructs step by step first
τ_ε (near τ), then α_ε (near α), then $\frac{1}{|\lambda_\varepsilon|}\tilde{G}_\varepsilon$ (near $\tilde{G}$) then
A_ε (near A) and finally $\frac{1}{|\lambda_\varepsilon|}\tilde{G}_{o,\varepsilon}$ (near $\tilde{G}_o$). All these bounds
are established in the next section where we then complete the
proof of Theorem 6.1.2.

6.3. The proof of Theorem 6.1.2 completed and concluding remarks

As indicated in step iv) of the equations which define the
operator K_ε we need to find a number τ_ε such that $\eta_\varepsilon(\tau_\varepsilon) = 0$,
where for a given function h_ε^u we have

$$\eta_\varepsilon(t) = -\lambda_\varepsilon + \frac{tx_{oo}}{\rho_h} \frac{\log t}{\log|\lambda_\varepsilon|} \cdot \frac{h_\varepsilon^{u'}(\frac{tx_{oo}^{-\alpha_h}}{\rho_h})}{1+h_\varepsilon^u(\frac{tx_{oo}^{-\alpha_h}}{\rho_h})} . \qquad (6.3.1)$$

Remember that in complete analogy we had to find for the case of the operator K a number τ such that $\eta(\tau) = 0$, where for a given function h^u we have

$$\eta(t) = 1 - \frac{tx_{oo}}{\rho_h} \log t - h^{u'}(\frac{tx_{oo}^{-\alpha}h}{\rho_h}) \ . \tag{6.3.2}$$

Lemma 6.3.1. Let $\{A, G_o\} \in E$ and let h^u be the function defined through (5.3.4), and corresponding to G_o and τ be the number defined through (5.3.5) and corresponding to h^u. There is a $\varepsilon_o > 0$ such that for $0 < \varepsilon < \varepsilon_o$ the function h^u_ε defined through (6.1.12), and corresponding to G_o and the number τ_ε defined through (6.1.13) and corresponding to h^u_ε satisfy Theorem 6.2.1 respectively the bound

$$|\tau_\varepsilon - \tau| < C \cdot \varepsilon (\log\varepsilon)^2 \tag{6.3.3}$$

where C is some constant independent of ε.

Proof : We need to prove the bound (6.3.3). First we note, that by Theorem 6.2.1 for $t \in I_\tau = [\tau - \Delta, \tau + \Delta]$, Δ small enough, ε fixed

$$h^u_\varepsilon(\frac{tx_{oo}^{-\alpha}h}{\rho_h}) = \varepsilon h^u(\frac{tx_{oo}^{-\alpha}h}{\rho_h}) + \varepsilon^2 (\log\varepsilon)^2 B(t) \ , \tag{6.3.4}$$

where we use the symbol B for functions which are continuous on I_τ. Similarly

$$\partial^i h^u_\varepsilon (\frac{tx_{oo}^{-\alpha}h}{\rho_h}) = \partial^i h^u (\frac{tx_{oo}^{-\alpha}h}{\rho_h}) + \varepsilon (\log\varepsilon)^2 B(t) \ , \quad i = 1, 2 \ ,$$

and therefore

$$\eta_\varepsilon(t) = 1 - \frac{x_{oo} \cdot t}{\rho_h} \cdot \log t \cdot h^{u'}(\frac{tx_{oo}^{-\alpha}h}{\rho_h}) + \varepsilon \cdot (\log \varepsilon)^2 B(t) \qquad (6.3.5)$$

and

$$\eta_\varepsilon'(t) = -\frac{x_{oo}}{\rho_h}((1 + \log t) h^{u'}(\frac{tx_{oo}^{-\alpha}h}{\rho_h}) + t \cdot \log t \cdot (\frac{x_{oo}}{\rho_h})$$

$$\cdot h^{u''}(\frac{tx_{oo}^{-\alpha}h}{\rho_h})) + \varepsilon (\log \varepsilon)^2 B(t) . \qquad (6.3.6)$$

Next note that from (6.3.5) it follows that for $z \in I_\tau$

$$\eta_\varepsilon(t) \Big|_{\varepsilon=0} = \eta(t)$$

and therefore $\eta_\varepsilon(\tau) \Big|_{\varepsilon=0} = 0.$ It follows that

$$|\eta_\varepsilon(\tau)| \leq C_1 \cdot \varepsilon (\log \varepsilon)^2$$

for some constant C_1 independent of $\varepsilon.$ Consider now the map

$$\hat{\eta}(t) = t - \eta_\varepsilon(t)/\eta'(\tau) ,$$

which is well defined on I_τ since $\eta'(\tau) \neq 0,$ which follows from the fact that the construction of τ in Section 5 did not lead to a domain error in our computer program. We have

$$|\hat{\eta}(\tau) - \tau| \leq |\frac{1}{\hat{\eta}'(\tau)}| C_1 \cdot \varepsilon (\log \varepsilon)^2$$

and furthermore for $t,$ with $|t - \tau| < 2C \cdot \varepsilon (\log \varepsilon)^2,$ we have

$$|\hat{\eta}'(t)| = |1 - \frac{1}{\hat{\eta}'(\tau)}(\eta'(\epsilon)|_{\epsilon=0} + C_2\epsilon(\log\epsilon)^2)| \leq C_3 \cdot \epsilon(\log\epsilon)^2 ,$$

for some constants C_2 , C_3 independent of ϵ. From these estimates follows the existence and uniqueness of a fixed point of the map $\hat{\eta}$ in the real interval $[\tau - \Delta_\epsilon, \tau + \Delta_\epsilon]$ with $\Delta_\epsilon = 2C \epsilon(\log\epsilon)^2$. Lemma 6.3.1 is proved.

To complete the proof of Theorem 6.1.2 is now an easy matter. One has just to plug in the bounds obtained for h_ϵ^u and τ_ϵ in the equations defining α_ϵ , $\tilde{G}_\epsilon^u$, $\tilde{A}_\epsilon$ and $\tilde{G}_{0,\epsilon}^u$ to prove the following lemma.

Lemma 6.3.2. Let h^u, h_ϵ^u , τ and τ_ϵ as in Lemma 6.3.1. Let furthermore α_ϵ , $\tilde{G}_\epsilon^u$, $\tilde{A}_\epsilon$ and $\tilde{G}_0^u$ be defined through (6.1.14)-(6.1.17) and α, $\tilde{G}^u$, $\tilde{A}$ and $\tilde{G}_0$ through (5.3.6)-(5.3.9). We have the bounds

$$|\alpha_\epsilon - \alpha| < C_1\epsilon(\log\epsilon)^2 ,$$

$$\|\frac{1}{|\lambda_\epsilon|}\tilde{G}_\epsilon^u - \tilde{G}_\epsilon^u\|_1 < C_2 \cdot \epsilon(\log\epsilon)^2 ,$$

$$|\tilde{A}_\epsilon - \tilde{A}| < C_3\epsilon(\log\epsilon)^2 ,$$

and

$$\|\frac{1}{|\lambda_\epsilon|}\tilde{G}_{0,\epsilon}^u - \tilde{G}_0^u\|_1 < C_4 \cdot \epsilon \cdot (\log\epsilon)^2 ,$$

for every $0 < \epsilon < \epsilon_0$, ϵ_0 small enough $C_1, \ldots, C_4$ are some constants independent of ϵ.

From this lemma Theorem 6.1.2 follows, using the representation (5.2.6), (5.2.7). In addition we see that the function $\theta_1(\varepsilon)$ of Theorem 6.1.2 is bounded by

$$\theta_1(\varepsilon) < C \cdot \varepsilon (\log \varepsilon)^2$$

for some constant C independent of ε.

In order to complete the discussion of our main theorem (Theorem 6.1.4) we should now give the proof of Theorem 6.1.3. The equations defining the tangent map DK_ε are however even more cumbersome to write down than those defining K_ε. Furthermore the constructions which are needed to prove this theorem are completely analogous to the ones we have already used to prove Theorem 6.1.2. Since nothing interesting can be learnt anymore by providing this proof we omit it.

Table 1 : The coefficients of the input sequence

-0.85968853396955692905E+00
-0.29059449620629290711E+00
-0.15174417755598792286E+00
 0.22638647193097692330E-01
 0.74083475460128505197E-01
 0.39725168852627255021E-01
-0.19973886824919282520E-02
-0.16175442075701530449E-01
-0.98616462511183513445E-02
-0.53229890455494194464E-03
 0.31937228633174328174E-02
 0.22077262441174758437E-02
 0.23917672724507693790E-03
-0.64874681595698913648E-03
-0.49544430929705854079E-03
-0.75185431661226706339E-04
 0.13495209399325987362E-03
 0.11428380852295876512E-03
 0.23809186949270167918E-04
-0.26530744877025793683E-04
-0.25644370831447788539E-04
-0.67078578779132047843E-05
 0.50457097838558192041E-05
 0.56357732415965729449E-05
 0.17333925271668879737E-05
-0.95954303801980270466E-06
-0.12457446956707538953E-05
-0.44302508454665014657E-06
 0.17324636661279615686E-06
 0.27264211811058243181E-06
 0.11075840898152552961E-06
-0.28801113886121651854E-07
-0.58686930898964126228E-07

-0.26822903104396884867E-07
 0.43559978221593031447E-08
 0.12568192757755704985E-07
 0.64106562058462712543E-08
-0.50982293192460505053E-09
-0.26681789781448034694E-08
-0.15148023789600780571E-08
 0.87217493770843296848E-11
 0.55839777325566223101E-09
 0.35240268814681883786E-09
 0.20562965928567149276E-10
-0.11555837583746763199E-09
-0.81089057647889765858E-10
-0.94362242207011309556E-11
 0.23609193222968358921E-10
 0.18488243238294830672E-10
 0.31570186053412605020E-11
-0.47396171777549740266E-11
-0.41716755120271856042E-11
-0.92660381921647879712E-12
 0.93295332035711965677E-12
 0.93249187538090002204E-12
 0.25321895919665351636E-12
-0.17924375100560300655E-12
-0.20664414590263894969E-12
-0.66173332370499578399E-13
 0.33307912659516605413E-13
 0.45379822331665222584E-13
 0.16743051167690877661E-13
-0.59054885405778608301E-14
-0.98749088257493206327E-14
-0.41332281546499489294E-14
 0.97326511114312175710E-15

Acknowledgements

The present work has extended over several years and we are particularly grateful to Henri Epstein for his continuing interest and support, and especially for bringing Ecalle's work to our attention. We have also profited from the stimulating atmosphere at the Institute for Mathematics and its Applications, Minneapolis and at the IHES in Bures-sur-Yvette, and we thank for their financial support. We also acknowledge financial support from the Fonds National Suisse, from the Institute for Advanced Study and from the Courant Institute.

Finally, we are indebted to Francine Nicole for her patient and excellent typing of our illegible manuscript.

References

[1] G. Benettin, C. Cercignani, L. Galgani and A. Giorgiolli, Lettere al Nuovo Cimento $\underline{29}$, 163-166 (1980).

[2] T. Bountis, Physica $\underline{3D}$, 577 (1981).

[3] M. Campanino, H. Epstein et D. Ruelle, On Feigenbaum's functional equation. Topology $\underline{21}$, 125-129 (1982).

[4] M. Campanino and H. Epstein, On the existence of Feigenbaum's fixed point. Comm. Math. Phys. $\underline{79}$, 261-302 (1981).

[5] P. Collet and J.-P. Eckmann, Properties of continuous maps of the interval to itself. In : Mathematical Problems in Theoretical Physics, Procesdings, Lausanne 1979. Berlin, Heidelberg, New York, Springer 1980.

[6] P. Collet and J.-P. Eckmann, Iterated maps on the interval as dynamical systems, Progress in Physics, Boston, Birkhäuser 1980.

[7] P. Collet, J.-P. Eckmann and H. Koch, Period doubling bifurcations for families of maps on $\mathbb{R}^n$, J. Stat Phys. $\underline{25}$, 1 (1980).

[8] P. Collet, J.-P. Eckmann and H. Koch, Physics $\underline{3D}$, 457-467 (1981).

[9] P. Collet, J.-P. Eckmann and O.E. Lanford III, Universal properties of maps on the interval, Comm. Math. Phys. $\underline{76}$, 211-254 (1980).

[10] P. Coullet and C. Tresser, CRAS Paris $\underline{287A}$, 577 (1978); J. Physique Colloque $\underline{C539}$, C5-25 (1978)

[11] B. Derrida, C. Itzykson and J.M. Luck, Oscillatory critical amplitudes in Hierarchical Models, Commun. Math. Phys. $\underline{94}$, 115-132 (1984).

[12] J. Ecalle, Publications Mathématiques d'Orsay No 67-7409.
Théorie des Invariantes Holomorphes.

[13] J. Ecalle, Publications Mathématiques d'Orsay No 81-05,06,07.
Les fonctions resurgentes (en trois parties) :
- Tome I   : les algèbres de fonctions résurgentes
- Tome II : les fonctions résurgentes appliquées à l'itéra-
             tion.
- Tome III : to appear.

[14] J.-P. Eckmann, H. Epstein and P. Wittwer, Fixed points of
Feigenbaum's type for the equation $f^p(\lambda x) = \lambda f(x)$, Commun.
Math. Phys. 93, 495-516 (1984).

[15] J.-P. Eckmann, H. Koch and P. Wittwer, Existence of a fixed
point of the doubling transformation for area-preserving maps
of the plane, Phys. Rev. A/R 26, 720 (1982).

[16] J.-P. Eckmann, H. Koch and P. Wittwer, A computer-assisted
proof of universality for area-preserving maps, Memoirs AMS 47,
289 (1984).

[17] J.-P. Eckmann and P. Wittwer, Sur un cas limite de l'équation
de Cvitanovic-Feigenbaum, C.R.Acad.Sc. Paris, t.299, No 4 (1984).

[18] H. Epstein and J. Lascoux, Analytic properties of the Feigenbaum
function, Comm. Math. Phys. 81, 437 (1981).

[19] M.J. Feigenbaum, Quantitative universality for a class of nonlinear transformations, J. Stat. Phys. 19, 25-52 (1978); 21,
669-706.

[20] M.J. Feigenbaum, The transition to apperiodic behaviour in
turbulent systems, Comm. Math. Phys. 77, 65-86 (1980).

[21] See [49].

[22] Fundamentals of numerical computation, Computing Supplementum 2, Springer-Verlag 1980, and references therein.

[23] J.M. Greene, R.S. MacKay, F. Vivaldi and M.J. Feigenbaum, Physics $\underline{3D}$, 468-486 (1981).

[24] J. Guckenheimer, Bifurcations of dynamical systems, CIME Lectures 1978.

[25] J. Guckenheimer and P. Holmes, Nonlinear Oscillations, dynamical systems and bifurcations of vector fields. Applied Mathematical Sciences $\underline{42}$, Springer 1983.

[26] P.R. Hauser, C. Tsallis and E.M.F. Curade, Phys. Rev. A (1984).

[27] R.H.G. Helleman, in : Fundamental Problems in Statistical Mechanics, Ed. by E.G.D. Cohen (North-Holland, Amsterdam, 165), 1980.

[28] E. Hille, Ordinary differential equations in the complex domain, J. Wiley, New York 1976.

[29] B. Hu and J.M. Mao, Period-doubling : Universality and critical-point order, Phys. Rev. $\underline{A25}$, 3259 (1982).

[30] T. Kato, Perturbation theory for linear operators, in : Die Grundlehren der mathematischen Wissenschaften in Einzeldarstellung, Band 132.

[31] O.E. Lanford, Smooth Transformations of intervals. Séminaire Bourbaki No 563, 1980/81, Lecture Notes in Mathematics Springer.

[32] O.E. Lanford III, A computer-assisted proof of the Feigenbaum conjectures, Bull. AMS, New Series $\underline{6}$, 127 (1984).

[33] J.-J. Loeffel, preprint, Lausanne (1976, 1977).

[34] B. Malgrange, Travaux d'Ecalle et de Martinet-Ramis sur les systèmes dynamiques. Séminaire Bourbaki 34e année, 1981/1982, No 582.

[35] R. Vilela Mendes, Critical point dependence of universality in maps of the interval, Phys. Lett. 84A, 1 (1981).

[36] R.E. Moore, Interval analysis, Prentice-hall, Series in automatic computation 1966, and Methods and applications of interval analysis, SIAM, Philadelphia 1979.

[37] G. Parisi, Asymptotic estimates in perturbation theory, Phys. Lett. B66, 167-169 (1977).

[38] G. t'Hooft, Borel summability of a four-dimensional field theory, Phys. Lett. 119B, 369-371 (1982).

[39] E. Brézin, J. Le Guillou and J. Zinn-Justin, Perturbation theory at large order. I. The ϕ^{2N} interaction, Phys. Rev. D15, 1544-1557 (1977).

[40] G. Roepstorff, Private communication.

[41] A. Voros, The return of the quartic oscillator. The complex WKB method, Ann. Inst. Henri Poincaré 211-338 (1983).

[42] G. Hardy, Divergent series (Oxford U. Press, 1949).

[43] G. Doetsch, Handbuch der Laplace-Transformation, Birkhäuser, Basel 1956.

[44] B. Simon, Large orders and summability of eigenvalue perturbation theory : A mathematical overview, Int. J. Quantum Chem. 21, 3-25 (1982).

[45] J.-P. Eckmann, J. Magnen and R. Sénéor, Decay properties and Borel summability for the Schwinger functions in $P(\varphi)_2$

theories, Commun. Math. Phys. $\underline{39}$, 251-271 (1975).

[46] J. Magnen and R. Sénéor, Phase space cell expansion and Borel summability for the Euclidean φ_3^4 theory, Commun. Math. Phys. $\underline{56}$, 237-276 (1977).

[47] S. Graffi and V. Grechi, Commun. Math. Phys. $\underline{62}$, 83 (1978).

[48] P. Collet and J.-P. Eckmann, A renormalization group analysis of the hierarchical model in statistical physics, Lecture Notes in Physics 74, Springer, Heidelberg (1978).

[49] J.-P. Françoise, Translation representations of scattering for rational maps, in Singularities and Dynamical Systems International conference in Creta, Heraklion 1983 (North-Holland Series in Maths).

[50] A. Sokal, An improvement of Watson's theorem on Borel summability, J. Math. Phys. $\underline{21}$, 261-263 (1980).

[51] R. de la Llave and O.E. Lanford III, Work in progress.

Communications in
Mathematical Physics

Members of the IAMP are entitled to reduced subscription to the journal.

Subscription information and sample copies are available from your bookseller or directly from Springer-Verlag, Journal Promotion Dept., P.O. Box 105 280, D-6900 Heidelberg, FRG

ISSN 0010-3616

Title No. 220

Chief Editor: **A. Jaffe**, Cambridge, MA

Editorial Board:

H. Araki, Kyoto	Mathematical methods with direct relevance to physics
M. E. Fisher, Ithaca, NY	Statistical physics
J. Fröhlich, Zürich	Equilibrium statistical mechanics
R. Haag, Hamburg	Conceptual structure
S. Hawking, Cambridge, England	Gravitation and field theories
O. Lanford, Bures-sur-Yvette	Dynamical systems
J. L. Lebowitz, New Brunswick, NJ	Nonequilibrium statistical mechanics
G. Mack, Hamburg	Quantum field theory and statistical mechanics
J. Mather, Princeton, NJ	Classical dynamical systems
L. Nirenberg, New York, NY	Partial differential equations
K. Osterwalder, Zürich	Constructive quantum field theory and general theory of quantized fields
G. Parisi, Roma	General theoretical physics
B. Simon, Pasadena, CA	Schrödinger operators and atomic physics
Ya. G. Sinai, Moscow	Statistical physics and dynamical systems
T. Spencer, New York, NY	Disordered systems
S. T. Yau, La Jolla, CA	Relativity; geometry and physics

Advisory Board: **M. F. Atiyah**, Oxford; **F. Hirzebruch,** Bonn; **G. 'tHooft**, Utrecht; **R. Schrieffer**, Santa Barbara, CA; **I. Singer**, Cambridge, MA; **C. N. Yang**, Stony Brook, NY

Springer-Verlag
Berlin
Heidelberg
New York
Tokyo

Communications in Mathematical Physics is a journal devoted to physics papers with mathematical content. The various topics cover a broad spectrum from classical to quantum physics; the individual editorial sections illustrate this scope.

Lecture Notes in Physics

Selected Issues from

Lecture Notes in Mathematics